THE NEW MEDICAL CONVERSATION

THE NEW MEDICAL CONVERSATION

Media, Patients, Doctors, and the Ethics of Scientific Communication

Dennis J. Mazur

ROWMAN & LITTLEFIELD PUBLISHERS, INC.

Lanham • Boulder • New York • Oxford

ROWMAN & LITTLEFIELD PUBLISHERS, INC.

Published in the United States of America
by Rowman & Littlefield Publishers, Inc.
A Member of the Rowman & Littlefield Publishing Group
4720 Boston Way, Lanham, Maryland 20706
www.rowmanlittlefield.com

P.O. Box 317, Oxford OX2 9RU, United Kingdom

British Library Cataloguing in Publication Information Available

Library of Congress Cataloging-in-Publication Data

Mazur, Dennis John.
    The new medical conversation : media, patients, doctors, and the
ethics of scientific communication / Dennis J. Mazur.
        p. ; cm.
Includes bibliographical references and index.
    ISBN 0-7425-2028-5 (cloth : alk. paper) — ISBN 0-7425-2029-3 (pbk. :
alk. paper)
    1. Communication in medicine. 2. Communication in science.
    [DNLM: 1. Communication. 2. Ethics, Medical. 3. Interprofessional
Relations. 4. Mass Media. 5. Professional-Patient Relations. W 50
M476n 2002] I. Title
    R118 .M39 2003
    610.69'6—dc21
                                                        2002005565

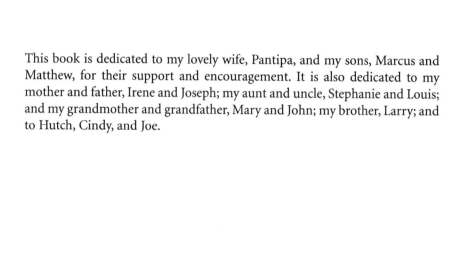

This book is dedicated to my lovely wife, Pantipa, and my sons, Marcus and Matthew, for their support and encouragement. It is also dedicated to my mother and father, Irene and Joseph; my aunt and uncle, Stephanie and Louis; and my grandmother and grandfather, Mary and John; my brother, Larry; and to Hutch, Cindy, and Joe.

Contents

Part II: How Information Reaches Patients

Part III: Communicating Risk-Benefit Information Today

Part IV: Communicating Risk-Benefit Information in the Future

Foreword

THIS BOOK'S FOCUS is on the short information message from the scientist, through the media, to the patient and to the clinician and other providers. It is about the impact of this message on the traditional patient–physician conversation, which becomes a new medical conversation with new roles for the consumer and patient, for the clinician and provider, for the scientist, and for the media people involved in the scientific research story and its communication.

What might be called the traditional patient–physician conversation had the physician as the primary source of information regarding what is needed for screening, diagnosis, and treatment. Here, the patient comes to the physician with questions and concerns, and asks the physician for his or her medical opinion. In this traditional patient–physician conversation, the physician is viewed as a patient advocate. The judicial system has taken this traditional view of the patient–physician relationship, for example, in Judge Spottswood Robinson's landmark federal case informed consent opinion in *Canterbury v. Spence*, which held that the physician's main task is information disclosure and the competent adult patient's main task is self-decision. The law also views the physician as a learned intermediary in the sense that manufacturers of medical products are duty-bound to inform physicians about risks associated with prescribed drugs and medical devices, and the physician then serves as a liaison between the product manufacturer and the patient.

Recent changes in the law have allowed product manufacturers, via advertising, to actually provide patients with information. While current direct-to-consumer advertising regulations support the statement at the end of each advertisement to the effect that the consumer needs to discuss the issues raised in the advertisement with his or her physician, the message contains elements of disclosures—about benefits and some risks related to their product—that previously had been part of the physician's domain of disclosure obligations. This abbreviated benefit-risk informational message is a prime example of what we shall call the short information message. And it is important to recognize that each short information message is characterized by an abbreviated risk component. Some might argue that this short information message contains both a short risk and a short benefit component, but the list of risks is always much longer than the list of benefits.

Yet, the direct-to-consumer advertising message is only one of an array of short information messages that confront consumers and patients. Most governmental health messages are also examples of short information messages. Indeed, most risk and benefit conversations in the traditional patient–physician conversation about risk and benefit related to a medical product are also examples of short information messages. This book is focused on the short information message and the challenges it presents to three groups: consumers and patients, who must interpret the short information message; clinicians, who must interpret the short information messages that patients receive from advertisers and from the government, among other groups; scientists, including those working for private industry, the government, and universities. This third set of challenges is particularly important, because the short information message has its origins in the scientific process that went into the development of a product and is now part of an advertised message, a governmental health message, or other short information message.

A fourth set of challenges faces the media. Through advertisements, news stories, and in-depth reviews of topics, the media play a pivotal role in the way medical science and medical research on humans is communicated to consumers, citizens, and patients.

This book was written for the following groups of citizens—consumer/patients, clinicians and other providers, scientists, media personnel, regulators, Congress and legislative bodies, and the judiciary. The book's purpose is to begin to make explicit the challenges we all face in making the best decisions we can for the health of our citizens and in presenting medical and scientific information to citizens and consumers.

Preface

M Y WORK OVER THE PAST TWENTY YEARS in information in clinical care and clinical research has been a major incentive for me to write this book. I work with young scientists, young investigators, and young physicians and physician researchers. Many of these young people were never approached with baseline issues related to clear communication in language nonscientists can understand. Many of these students never encountered the notion of *informed consent* in clinical care or clinical research. Now, as fellows and junior faculty, the young clinicians and young researchers are quite perplexed. When I review their manuscripts and their informed consent forms, I often make many recommendations. Many of these recommendations are in the form of suggestions for more disclosure of risk information to patients and better discussions of the risks their patients may encounter in clinical care or in a clinical trial. I help them to get out of the habit of formulating short information messages. These short messages often hide more than they disclose regarding the nature of the clinical interventions or research interventions. I help them to begin to develop sound and coherent approaches to the *full disclosure* of risk and uncertainty to their patients who want and need this information to help them understand the situation they face and the decisions that need to be made in clinical care and clinical research.

Yet the patient–physician and the human subject–principal investigator relationships are just the tip of the iceberg in terms of the communication of scientific information to citizens. The media working on communication of scientific discovery put new requirements and a new form to the discussions held by patients and physicians in clinical care. I argue that medicine and science have created a "new medical conversation" in the patient–physician relationship. And this conversation places a new obligation for clear communication on all of us—media, scientists, doctors, and patients—that perhaps has never been seen before as necessary within a citizenry.

Finally, with a new interest in federal regulations in clinical research on human subjects, the goals of clear communication and clear informing need to be met much more quickly than at any time before. In human genetics, there are many new issues that involve more than simple consideration of risk by an individual who may face employability and insurability consequences. Yet many of the issues that currently hit the media are issues of lack of attention to basic regulations, basic science, and basic informing of patients about the fact that severe adverse outcomes at low probabilities do in fact occur. These issues of what I call "basic communication for the goals of basic informedness" have perplexed medical science, doctors, and patients since at least the late 1700s. Yet issues of what constitutes clear communication and clear informing of individuals who are nonscientists are not yet resolved and need even closer attention in 2002 and beyond. Why? The economic constraint on clinical care and clinical research and in patient–physician decision making has forced clinicians and researchers to pay much closer attention to becoming better communicators in order to help their patients understand the decisions that confront them in clinical care.

This book is written for future media people, future scientists, future physicians, and future patients. I wrote this book to help all groups better understand the issues that continue to confront us on a day-to-day basis as we strive to communicate information clearly. Clear communication is the goal we need to strive for in order to better serve the recipients of our information messages. They should be better decision makers and decision trackers in their own lives for having received our messages.

In writing this book, I hoped to approach students who are still at a distance from the actual application and development of concepts, projects, programs, and research in the areas of scientific information and its interface with many disciplines. These disciplines include the sciences, social sciences, communication sciences, journalism, media sciences, philosophy, public health, and the medical sciences: those who will eventually find careers in fields related to science and medicine. I wrote this book in the hope that students will have the time now that they will not have when they are actively pursuing their respective careers, to think about how they are going to be involved in the communication of scientific information in a rational and an ethical manner.

Yet I believe this book will also be of interest to other students, particularly consumers and patients, as they develop their own perspectives on how they will evaluate and incorporate scientific information, with its social nuances, into their own medical and health care decision making. Here, the consumer as the receiver or recipient of the health care message is perhaps the best evaluator of the message. But I believe the evaluator of the message must be able to understand and use the concepts this book provides in sorting out the wheat from the chaff in decision making.

One of the key objectives of this book is to make a clear distinction between the *advertising message* and the *information message* for decision making. Physicians need this distinction to help them communicate with their patients more effectively. Principal investigators need to focus on this distinction to help them communicate with human subjects regarding the realities of participation in clinical scientific studies and the chances of severe adverse outcomes occurring even when the highest levels of protection are being pursued. Informed consent in clinical care, for the most part, relates to informing patients about the risks, benefits, and alternatives of procedure that are more or less standardized in the practice of medicine. Clinical research involves the testing of new scientific hypotheses and extensions of sciences into new dimensions of care.

Finally, I believe this book will be of use to the physicians, researchers, investigators, and administrators at all levels and in all work environments aimed at developing decision support for patients within local, regional, national, and international health and medical care systems. The responsibilities of the developers of decision support are perhaps the greatest of all, because they are most probably the furthest removed from the face-to-face interaction that a patient can engage in with a doctor. In many decision support systems, the patient interacts directly with a computer program. This type of interaction is a great distance away from the traditional patient–physician relationship where there is eye-to-eye contact between the patient and the physician. Thus, the development of a computerized interactive decision support system needs even greater scrutiny than less structured patient–physician environments. Interactive systems without opportunities for direct questioning of another human being force us to consider what is to count as an optimal or a suboptimal interaction. What is successful communication and what is a failure? What criteria are to be used? What are the gains and what are the losses? And who is doing the judging?

Undergraduates need a survey of the different types of decision-making processes and the challenges faced by patients and physicians in grappling with information in all of its forms and all of the ways it is presented. Undergraduates from various fields can benefit from this book because of its emphasis on the basic dialogue between the patient and physician, a dialogue in which everyone will have to select the role he or she wants to play in the relationship. What kind of physician does an individual want as a care provider? What kinds of information will future patients want? What kinds of information will future scientists, future medical researchers, and future government administrators provide to consumers, patients, and citizens? Information must be understood in terms of the three decision-making scenarios that individuals can opt for in their relationships with the physicians caring for them:

- patients making the decision themselves;
- patients sharing the decision making with their physicians;
- patients tracking the decisions made by their physicians.

For graduate students, this book offers a survey of the key theoretical issues in the development of information and decision support in the patient-physician dialogue. It will facilitate better theory development, better research, and better approaches to decision making, based on specific considerations:

1. the availability of solid information that can be used by patients and physicians, as opposed to information that is problematic in terms of its sources;
2. the ways in which information is presented to both patient and physician audiences;
3. the nature of the message conveyed.

Graduate students in the social sciences and philosophy who see themselves as future researchers in this field may be interested not simply in further clarifying information in the patient-physician relationship but in going on to develop decision systems designed to help patients and physicians sort out, prioritize, and systematically deal with the information that is available to them through computerized decision support.

Acknowledgments

I THANK THE FOLLOWING COLLEAGUES for their help and support over time: David Hickam, James Reuler, Gordon Noel, Larry Oresick, William Holden, Stephen Campbell, Ed Murphy, Mark Helfand, John Kendall, Leslie Burger, Ted Galey, Michael Davey, John Mather, Susan Tolle, Linda Humphrey, Martha Gerrity, Heidi Nelson, Beverly Jefferson, Thomas Cooney, Karen Rasmussen, David Nardone, Donald Girard, David Smith, Kusum Kumar, Linda Lucas, Som Saha, Greg Magarian, Mary Glanville, Nancy Fletcher, May Jessup, Chris Theonnes, Michael Sturges, and Andrée Coracy.

In addition, I am grateful for the support of my colleagues in each of the following areas: In philosophy, Julius Moravcsik, Jaakko Hintikka, and Patrick Suppes; in biostatistics, Jodi Lapidus and Katherine Riley; in the education of educators, Kelley Skeff and Robert Waterman; in law, Jon Merz; in cognitive psychology, Amos Tversky; in medical decision making, Harold Sox, Keith Marton, Stephen Pauker, Arthur Elstein, J. Sanford Schwartz, J. Robert Beck, Lee Lusted, Eugene Sanger, John Eisenberg, Denny Fryback, Robert Nease, Doug Owens, Ben Littenberg, Frank Sonnenberg, Mark Eckman, Nananda Col, Adrian Edwards, Hillary Lewellyn-Thomas, H. J. Sutherland, and the members of the Society for Medical Decision Making and the European Society for Medical Decision Making, especially Wilfried Lorenz of Marburg, Germany; in the area of evidence, Mark Helfand, Linda Humphrey, Heidi Nelson, Martha Gerrity, and the research team at the Evidence-Based Practice Center at the Oregon Health and Science University and the Project Development Team for the Evidence-Based Practice Center's work for the preparation of the third edition of the U.S. Preventive Service's

Task Force report; in the information sciences, William Hersh; in the protection of human subjects, Stephen Hefeneider, Richard Jones, Mabel Gearhart, Max Metcalf, Ruth Whitham, William Wickham, John McDermott, Theresa Harvath, William Hoffman, William Tuttle, Wayne Clark, Sue Millar, Vickie Vonderoe, Ron Brown, Richard Yeager, Sola Whitehead, and Jodi Pass.

I have benefited in my own thought and development from the vigorous debates, careful thinking, and penetrating comments and insights of the members of the Research-in-Progress seminars. These seminars are held jointly by the Section of General Medicine at the Portland Veterans Medical Affairs Center (VAMC) and the Division of General Medicine at OHSU. I thank Sandra Brayson, Kathy Jordon, and Mara Wilhelm for performing library searches of the peer-reviewed medical literature; Carol Franks for editing and proofreading; and colleagues I have long read: Paul Slovic, Baruch Fischhoff, and Jay Katz.

I also thank Dean Birkenkamp, Alden Perkins and Alison Sullenberger for their tremendous help and support through the completion of this project.

I

Key Points Needed
for Better Communication

1

Introduction

A NEW ERA IS UPON THE SCIENTIST and the physician. It is based on the accurate communication of information. There are many unanswered questions regarding what it means to successfully communicate a message to a medical patient or human research subject. The potential problems in communication exist not only among physicians and scientists but also between physician and patient and between principal investigator and human subject. Undergraduate and graduate students need to prepare themselves conceptually for the issues that will confront them when they leave school and training and enter careers that involve patient care or scientific research on human subjects. Such research is conducted at many universities and in medical centers in the United States. Yet this arena is changing with the introduction of heightened protections for the potential human subjects in a clinical trial.

Issues of informed consent in relation to drugs and devices have been part of federal regulation since the mid-twentieth century. The term *informed consent* in clinical care has been used in judicial discourse since 1957, but the concept continues to be hotly debated in terms of what actually is to be communicated to the patient about risk and benefit. The challenges are even greater in clinical research, where patients may face only risks and may accrue no benefits during their participation. Yet these challenges have existed throughout time, as physicians have conducted invasive medical interventions and researchers have conducted moderate- and high-risk studies on patients.

Today, however, there is an increasing focus on research involving the human genome and genetic research involving DNA. Here the old rules of informed consent in research—still themselves under scrutiny and often difficult to apply in informed consent documents—do not apply, because the research subject is no longer a single individual. That is, the individual considering his or her consent to donate genetic material or unique protein material for storage and use in future research is not the only one at risk. Also at risk are his or her present blood relatives and indeed the future generations of the family line. The issues of what should count as a successful communication are just beginning to be addressed and are slow to be clarified in federal regulation. Regulations are guidelines. The real decision making, in terms of applying these guidelines in the construction of informed consent documents, is only beginning to come to the attention of institutional review boards already overworked with current research study evaluation and clarification of informed consent forms.

I wrote this book to be used in a new type of class for undergraduates and graduate students in the sciences, to better train our future scientists, physicians, and researchers on the problems that now exist regarding the communication of information. These problems have existed since the first patient underwent an invasive (drug or device) intervention on the recommendation of his or her physician. Now in genomic research, communication issues involve risks and benefits for parties extending beyond both the individual to his or her family and future generations. By developing a broad-based approach to understanding the present parameters of communication about risk and benefit from key areas in law, ethics, and the social sciences, I hope to bring this topic into the training of undergraduates and graduate students much earlier, when they have more time to consider the issues.

From the perspective of the social scientist, the issues of the new medical conversation, with attention now to the ethics of scientific communication, will begin to be felt in the number of parameters that impact the patient–physician or the principal investigator–subject dialogue. From the philosopher's perspective, the issues of the new medical conversation will force more attention on how information, particularly scientific information, is communicated in each encounter. In addition, ethicists must be attentive to how they interrelate with the scientific information that is part and parcel of the patient's decision.

I believe that better trained and better educated undergraduates and graduates will bring new insights to communication challenges. This book was written so that students would have the opportunity to think these issues through earlier in their careers.

2

Media, Science, Doctors, and Patients

W^E WILL FOCUS ON FOUR VIEWS OF THE WORLD: the view of the media, the view of science, and the views of doctors and patients trying to deal with scientific information coming at all of us from many directions. Science becomes the "hook" to draw the consumer's attention to what is often a short, non–scientifically presented advertising message. Much of the information comes at us in "sets" or hooks to catch our attention, but little instruction is given to consumers as to how to unset the hooks once their attention has been captured.

Using Scientific Information to "Set the Hook"

It is fair to say that the notion of "setting the hook," while derived in part from basic product advertising approaches, has also been taken up and used by contemporary educators, for example, teachers teaching teachers to teach. Barbara Gross Davis gives the following points for opening a lecture:

- Avoid a cold start.
- Minimize nervousness.
- Grab students' attention with your opening.

Davis recommends that the lecturer "open with a provocative question, startling statement, unusual analogy, striking example, personal anecdote, dramatic contrast, powerful quote, short questionnaire, demonstration, or mention of a recent news event."[1]

Davis's first example focuses on the use of numbers in a sociology lecture:

> How many people would you guess are sent to prison each week in the state of California? Raise your hand if your think 50 people or fewer. How about 51 to 100? 101 to 150? Over 150? (Pause) In fact, over 250 people are placed in custody every week.[2]

When Should a "Hook" Be Considered Inappropriate, Unfair, or Deceptive?

Medical tort law forbids clinicians from engaging in fraud and misrepresentation in the patient–physician relationship. Here Fay Rozovsky argues:

> Misrepresentation of material information or fraudulent concealment can invalidate consent to treatment. To establish such a claim, the litigant must prove that the defendant misrepresented or suppressed material details and possessed any knowledge of falsity. It must be clear that there was an intent to induce reliance on the misrepresented information and that there was in fact actual and justifiable reliance on the information. Finally, there must be proof of resulting damages.[3]

Yet, the courts rely heavily on interpretations of "material information." How do you establish that someone "intentionally misrepresented or suppressed information"? How do you establish that there was "an intent to induce reliance on misrepresented information"?

Noting the problems with communications between physicians and patients in situations involving prognosis (prognostic information)—for example, how long will the patient survive?—Peter Ubel argues that there is a conception regarding truth in this type of circumstance that carries with it the notion of optimism. Ubel argues that "the truth we communicate to patients should help them prepare for the worst while allowing them to hope for the best."[4]

Most of the examples we will be discussing in the book come from the clinical medical sciences, where there is an obligation for principal investigators to provide human subjects as much information as possible. However, many groups of investigators in the social sciences face dilemmas. These investigators have explicitly discussed the issue of deception of human subjects in conjunction with such research.

Sears, Pelpau, and Taylor note the problems social psychologists face when approaching a potential research subject: "Informed consent means that the subject must voluntarily agree to participate without any coercion. . . . The researcher has an obligation to tell potential subjects as much as possible about the study before asking them to participate."[5] While agreeing that the requirements sound reasonable, the authors note that "informed consent . . . can sometimes create problems for social psychologists. . . . [It] may be important not to tell subjects the true purpose of the study, to avoid biasing their responses. Even in the simplest research, subjects are rarely told the specific hypotheses that are being tested." They continue:

> Some people believe that deception of any kind is unethical in psychological research. They think it demeans the subjects and should never be used. A more moderate position endorsed by most research psychologists is that deception should not be used if at all possible, or used only after considering its possible harmful effects. Subjects, however, should always be volunteers. Perhaps they need not be told everything that will happen, but they should know that they are in an experiment and should have freely given their permission. In other words, only someone who has given informed consent or consent based on trust should be exposed to potentially distressing conditions.[6]

We can now ask to what extent the set or hook—for example, using only a part of available scientific information related to a medical product or device—should be considered, at one extreme, unfair, and at the other, deceptive. The simple answer is that a set or hook is fair when there is a follow-through on the scientific information presented. I argue that it is important for people to see if there is a follow-through, to learn how to unset the hook, and then to evaluate the information presented.

The big question regarding the use of hooks is determining whether there is a follow-through on the scientific information. One can point to the drug advertisements on television that mention two to three common side effects and then direct consumers to their physicians for some sort of more elaborate explanation or follow-through. Some advertisers may provide a site on the World Wide Web with additional scientific points, but rarely, in my experience as a consumer, have I seen what I would call a full scientific follow-through on the brief elements of information that are presented in advertisements.

In the introductory lecture for the basic biostatistics course offered by their university, students saw the following statement flashed on the projector screen: "1.6 percent of women get breast cancer."

The phrase was used to set the hook for an upcoming discussion regarding what is meant by that 1.6 percent and the problems with the number's use and its interpretation. One would hope that following this set would be much more expansive clarifying information that would flesh out the science implied by the figure.

Some students simply jotted the statement down in their notebooks. Others committed it to memory. Still others looked puzzled. One whispered, "Isn't that number too low?" Another replied, "Too low for what?" Two students agreed that the risk of a woman developing breast cancer in the United States was one out of eight or one out of nine. Yet, the majority of the audience was content with the number. No one was willing to ask a question.

The instructor started the lecture: "You have all had some time now to examine the number on the screen since you entered the lecture hall. What do you think about that number?"

The following questions were generated by the audience:

- Which country did this data come from?
- Which age groups of women does this number apply to? Or does the number apply to all women everywhere?
- How accurate is the number? Is the number an estimate or best guess?
- If an estimate, who is making the estimate? How qualified is the estimator?
- Is the estimate too high, too low, or right on target?

Other questions pertain to the audience members themselves:

- Is the lecturer's point about the number a point about other women and not me? Or is the point about me?
- How should I react to the number?

During the course or in the lecture, the audience may ask further questions, or, if not asked directly, the lecturer may pose additional questions:

- How do you know the number is valid?
- What type of evidence is behind the number?

Then the lecturer may really start digging into the subject: "How large was the population being studied? That is, is it 1.6 percent of 50 women, 500 women, or 5,000 women?" Then the lecturer might ask the audience to consider that the reliability or significance of that 1.6 percent is dependent on how large the study population was. The lecturer might ask, "Were the women studied randomly selected from the population at large?" Here, a randomly selected population would yield more scientifically reliable results than a non-randomly selected population. Then the lecturer might proceed by asking, "What does it mean to randomize?"

Yet by the end of the lecture no one had brought up the issue that they were not certain where the number came from. The next day, the teacher noted in a comment to a student that it was from NHANES. But the teacher gave no further information related to the number or what NHANES meant. The teacher did note that the number of women studied was 1,628.

After asking some of the other students in the lecture regarding the number and not getting a satisfactory response, one student decided to pursue the information issue in a systematic fashion. The student decided to use two approaches. Approach 1 was based on the scientific literature related to breast cancer. Approach 2—not to be pursued if Approach 1 yielded the answer—was to check with experts in the field.

Approach 1: Systematically searching the scientific literature from a student's perspective

The student went home, powered on the computer, and searched PubMed without real success. The student then used a search engine to obtain a set of websites and documents related to NHANES. The student found that NHANES referred to the National Health and Nutrition Evaluation Survey, a survey conducted by the National Center for Health Statistics (NCHS), Centers for Disease Control and Prevention. The student found references for documents related to three studies, with the names, NHANES I, II, and III.

But at this point the student, unassisted, got lost in attempting to move the project any further, and has not been able to find the answer to the question within the contents provided on the Center for Disease Control website. Thus, the student could not verify that the statistic cited by the lecturer was in fact found in the report of NHANES I, that the statistic was real, or in what sense it was real.

Although some information was gained in the search, the student still did not understand why the 1.6 percent figure was used. Was it a historical point, that the number that came out of NHANES I would be different today? Or was it some other point? The student needed to ask the lecturer.

Approach 2: Asking experts in the field

The student decided to go to the medical school and speak with residents. The secretary allowed the student to enter the clinic area to ask a question. The medical residents were very busy with their patients, but the student got the following answer:

"I'm not familiar with NHANES, but I believe women in the U.S. have a one-in-eight chance of developing breast cancer."

Another resident overhearing the discussion could shed no light on NHANES but offered up the following: "I believe that the chance for a woman developing breast cancer is one in nine." A third resident directed the student to the attending physician. The student asked if the attending physician was aware of a one-in-eight or one-in-nine chance of women developing breast cancer, and what this actually meant. The attending physician replied, "The number I carry around in my head is 10 percent. I believe this represents the chance of a woman in the United States developing breast cancer at some time in her life." However, he pointed to the following caveats:

- Age influences the chance of a woman in the United States developing breast cancer, with older women having a higher risk.
- There are suggestions in the peer-reviewed medical literature that certain prescribed medications, for example, hormone replacement therapy, may increase a woman's chance of developing breast cancer.
- The presence of breast cancer in a woman's family increases her chances.
- However, according to the report of the U.S. Preventive Services Task Force, second edition (1996), a woman's estimated lifetime risk of dying from breast cancer is 3.6 percent.[7]

The attending physician paused and looked at the puzzlement in the student's eyes.

"But wait, what I'm telling you now is what I'm holding in my memory about this issue. Just wait a second. . . ."

The attending physician eyed the bookshelf and spotted a copy of *Guide to Clinical Preventive Services*, second edition. He walked over, thumbing through the book as he went back to the student.

"Here, take a look at chapter 7, 'Screening for Breast Cancer.' This is what the U.S. Preventive Services Task Force said in 1996."

Important risk factors for breast cancer include female gender, residence in North America or northern Europe, and older age. In American women, the annual incidence of breast cancer increases with age: 127 cases/100,000 for women aged 40–44; 229/100,000 for women aged 50–54; 348/100,000 for women aged 60–64; and 450/100,000 for women aged 70–74. The risk for a woman with a family history of breast cancer in a first-degree relative is increased about 2–3-fold, and for women under 50 it is highest when the relative had premenopausally diagnosed breast cancer. Women with previous breast cancer or carcinoma in-situ and women with atypical hyperplasia on breast biopsy are also at significantly increased risk. Other factors associated with increased breast cancer risk include a history of proliferative breast lesions without atypia on breast biopsy, late age at first pregnancy, nulliparity, high socioeconomic status, and a history of exposure to high-dose radiation. Associations between breast cancer and oral contraceptives, long-term estrogen replacement therapy, obesity, and a diet high in fat have been suggested, but causal relationships have not been established.[8]

The attending physician approached the student.

"Here, I signed this out of our small library here under my name. Take it home and I'll see you back here next week and we can talk about it further."

The student thanked the attending physician and promised to be back next week.

"Oh, by the way," the attending physician added, "those numbers might be adjusted because there is a new task force report out in a year or so. You'll have to get used to it. These numbers need to be updated all the time."

The precise location of this number from NHANES has not been found to this date by those students from the class who elected to pursue the search of NHANES.

Let us now look at another statement: "one out of eight women get breast cancer." Consider what happens when this statement is used in an advertisement for a product or service but is not developed further. Here is a short information message with an ambiguous element that appears to be scientific information. But it is a fragment of scientific information without any further discussion or explanation as to the origins of the information in terms of what science was done to get that result or what science and biostatistics can help us to interpret that result.

Before we were talking about the topic of a course or at least of an hourlong lecture about the statement and its ramifications. However, the advertiser's strength (and our weakness as consumers) is that the advertiser's message may be quite short, and this may be the only semblance of a statement remotely reminiscent of science that we are going to get. These points or statements flash across the television screen in advertisements about a medical product (a drug or device). In the first case, one anticipated a full course or lecture to get at the answers to this set or hook. However, where does the consumer go to for enough information to answer the question relative to her own particular circumstance? What if the numbers instead were about the chance of a man developing prostate cancer? The same questions would still apply regarding what those numbers mean, and our final question also would apply: Where should the man go to get the answers he needs?

Setting the Hook and Unsetting the Hook

Setting the hook is actively taught as an attention-getting device in a variety of arenas: from advertising to lecture courses about scientific topics, to training for medical students, to the continuous training clinicians receive over their careers to keep up-to-date within an ever-increasing scientific domain. What is not done as routinely is to learn how to unset this hook once it is placed and to evaluate the information that follows the placement of the hook.

The burgeoning of the sciences, in particular the medical sciences, and the consequent review and analysis of these sciences in the media, have helped create what I consider to be a new medical conversation. Traditionally, the medical conversation has involved a two-way information exchange between a patient and a physician. However, today, science and media—through the communication of information to the public in news stories, commentaries, and advertisements containing bits and pieces of scientific information—have put different sources of scientific information into the hands of patients. And patients, in turn, bring new questions and requests to their dialogue with their doctors.

However, there are costs to these new sources of information. There are economic costs to society in terms of increased requests for newly marketed prescription drugs. There are emotional costs to the patient, who may be a nonscientist perplexed by the information and who may request guidance from a physician. And there are costs to the patient–physician relationship because of new time pressures placed on the physician as the interpreter of scientific information and scientific perspectives. In fact, the physician may not be sufficiently aware of new information. Scientific information itself needs scrutiny, to see which hypothesis out of a set of hypotheses bears true over time.

The physician, who has long been considered a learned intermediary within the law in relation to medical products (drugs and devices) that require prescription, now becomes a new type of intermediary: a scientific intermediary asked to interpret, explain, and put in perspective a snapshot of medical scientific information. At the same time, a drug may be so new that it has not had the intense scrutiny needed to see how it performs in the much more medically complex environment of the general population.

The problem is that not all patients and physicians want to wait one to five years for the risks and benefits of a drug to be reported in the peer-reviewed medical literature. And indeed these new risks and benefits will not be found until the drug is used in 20,000 patients. When a patient comes to a physician with a question or request about a newly marketed, newly advertised drug, the physician is asked to explain, interpret, and formulate an opinion on new science. But will there be a reproduction of beneficial results when the drug is used in a medically more complex set of patients? Will new risks be uncovered? Until a drug is studied in patients with medical conditions of various degrees of complexity over time, the jury is still out on whether the drug is safe and efficacious.

The balancing that is needed between the bits and pieces of scientific information provided in the media and the explanation, interpretation, and formulation of an opinion by a physician caring for that patient becomes the new medical conversation. This conversation is in need of an ethics of scientific communication. This book describes for the scientist, the nonscientist, the physician, and the patient the different perspectives that need to be considered in the development of such an ethics.

We will use the term doctors in this book mainly to refer to clinicians who provide medical care for patients. But it is important to recognize that the doctor may also be a media person, such as a medical correspondent, a medical commentator, or even an advertiser for a medical product. The doctor also could be a physician–scientist who spends all of his or her time in the lab and never sees a patient for clinical care, or the doctor could be a clinician–scientist who straddles the borders of clinical medicine and clinical research. The physician–scientist could work for the government, for a university, for a product manufacturer, or for an advertising firm focusing on drug advertisements for physicians. Finally, the doctor could be an entrepreneur in any of the above areas.

In clinical care, doctors are the prescribers of medical products (drugs and devices) and they are the physicians and surgeons who install these devices in patients who need them. As a physician, the doctor serves as a learned intermediary between the patient and the product manufacturer. The product manufacturer has duties to warn the physician about medical products, and the physician, in turn, must use these warnings in his or her judgments and communications with patients.

Physicians are also responsible for sorting out the product–practitioner dilemma. This dilemma exists with many medical products, and it involves the sorting out of adverse outcomes that occur with products—outcomes that may be due to the product or may be due to how the practitioner is using the product. Here, a practitioner's inexperience with the use of a regulated product or the off-label use of a product that is approved for another purpose may be a reason for an adverse outcome in a patient.

We will use the term patients within this book to refer to individuals as medical consumers. But, more importantly, competent adult patients are decision makers regarding what is to be done with their bodies. Competent adult patients may decide to make decisions for themselves, to allow the doctor or a significant other in their lives to make decisions on their behalf, or to share decision making in some way with their doctor. But the patient may also want to engage in nonmedical care and to use unregulated and untested products and devices that are advertised on television, on radio, and in newspapers, magazines, and journals across the United States. The doctor as clinician must ask the patient what dietary and herbal supplements the patient may be taking because of possible interactions with prescribed drugs. Here, physicians make inquiries into areas that the patient may not want or know about, and the physicians, in turn, must educate themselves in nonmedical matters about which little may be known.

New exploration into products that are unregulated and so far not systematically and rigorously studied may begin because of adverse outcomes that patients experience. Adverse outcomes could occur when patients add these agents to prescribed drugs they are already taking or undergo surgeries or other medical interventions when the physician does not know that these unprescribed medical agents are being taken.

Here, the initial observations and reports of clinicians and surgeons begin to lead to systematic and rigorous study of key topics in clinical medicine, now beyond the scope of regulated medicine and including unregulated substances. Yet, it is important to recognize that this approach of systematic science and observation and recording can become part of the following sequence of events: First, observation and monitoring become part of regular monitoring practices at medical institutions. Then, over time, analysis of the observational and monitoring data begins and can become incorporated into an overall approach to patient safety within clinical medicine and its important interrelationships with clinical pharmacology and medical product development.

This book is about the interface between media, science, doctors, and patients and the issues they face when aiming for patient safety in the context of freedom of choice for competent adult patients. Patient safety in the context of medical sciences is the focus of our attempt to develop an ethics of communication connecting media, medicine, science, doctors, and patients. While patient safety should be of utmost importance to all, new approaches, such as direct-to-consumer advertising of prescribed drugs, put an added burden on physicians as learned intermediaries. Now, physicians have new sets of key information that need to be communicated to and discussed with patients. This type of advertising pressure adds to already-full agendas for patient–physician conversation and discussion. This new medical conversation must be understood in terms of an ethics of communication that involves not only patients and physicians, but the media and scientists as well. This book is about how to optimize the role of all parties in terms of successful communication that helps patients understand the information that they need. In particular, the new medical conversation focuses on the communication of scientific evidence through the media and advertisements and on the impact of such communication on the patient–physician conversation.

Notes

1. B. G. Davis, *Tools for Teaching* (San Francisco: Jossey-Bass, 1993), 112.
2. Davis, *Tools for Teaching*, 112.
3. F. A. Rozovsky, *Consent to Treatment: A Practical Guide*, 3d ed. (Gaithersburg, Md.: Aspen, 2000), 1:13–1:16, quoted from p. 1:13.

4. P. A. Ubel, "Truth in the Most Optimistic Way," *Annals of Internal Medicine* 134, no. 12 (June 19, 2001): 1142–43, quoted from p. 1143.

5. D. O. Sears, L. A. Pelpau, and S. E. Taylor, *Social Psychology*, 7th ed. (Englewood Cliffs, N.J.: Prentice Hall,1991), 31.

6. Sears, Pelpau, and Taylor, *Social Psychology*, 32.

7. U.S. Preventive Services Task Force, *Guide to Clinical Preventive Services: Report of the U.S. Preventive Services Task Force*, 2d ed. (Baltimore: Williams & Wilkins, 1996), 73.

8. U.S. Preventive Services Task Force, *Guide to Clinical Preventive Services*, 73–74.

3

Basic Terms

W̲E WILL NOW REVIEW THE BASIC TERMS to be used in this book and discuss the interrelationships of terms.

The information types we will focus on are descriptions and estimates of chance (probability) and the level of uncertainty in these estimates. We will be interested in measurements and the accuracy of those measurements. We will be interested in research studies and what is communicated about them (their purpose, methodology, results, and various interpretations of results) to physicians, scientists, and consumers as well as to patients. We will be particularly interested in the translation process from the published medical and scientific literature to the consumer and patient.

We will focus on *scientific information*, particularly information derived from the medical and the clinical sciences, biochemistry, chemistry, and molecular genetics as it relates to research and the products of research—drugs and medical devices. In the cases of drugs and devices, the patient's physician is legally in the position of the *learned intermediary*. The physician stands between the patient and the decision on whether the drug is to be prescribed in the patient's case, in those cases where the drug is regulated by the Food and Drug Administration and where the patient can obtain the drug only with a physician's prescription.

Scientific Descriptions

Scientific descriptions are descriptions of medical interventions and their risks and benefits. These include the physical and physiological descriptions of the medical interventions in relation to drugs and medical devices as well as the descriptions of the benefits and risks of these drugs and devices. These include descriptions of *benefits, risks,* and *alternatives* as well as *temporal perspectives* regarding whether the decision must be made immediately or can be delayed, as well as the consequences of delay, if any.

We will also be interested in various *frameworks for evaluating scientific information* by scientists and nonscientists. Whereas scientists can review the medical and scientific literatures and discuss their interpretation of results with colleagues, the nonscientist may not have these luxuries. The nonscientist is thus presented with a shortened or abbreviated medical or scientific description of a medical drug or device.

In terms of scientific descriptions for nonscientists, we will focus on three areas:

- The translation of scientific descriptions into what the judicial system has called *lay language,* that is, the nontechnical language of consumers, patients, and others who are nonscientists.
- The nonscientific uses of these descriptions, for example, in advertising drugs and medical devices.
- How the subsets of scientific information are selected in various activities where scientific description is used for scientific and nonscientific ends.

We will be interested in written and spoken descriptions of the nature of medical interventions, the mechanism of action of drugs, and the way devices work. We will be interested in risks and benefits associated with drugs and devices, with success rates and failure rates in the initial populations in which they were studied and in the general population once marketed. We will need to confront issues of consumer resistance to numerical expressions of probability.

There are at least two basic types of scientific description. The first type involves written or verbal descriptions of hypothesis, study purpose, procedures, events, and processes. The second type involves numbers and numerical estimates. Each type of scientific description must account for two levels of description, one scientific and technical, the other also scientific but translated into language that can be understood by nonscientists.

1. Scientific Descriptions of "Hypothesis, Study Purpose, Procedures, Events, and Processes"

The first type of scientific description involves descriptions of scientific hypotheses that are being tested or that have been tested and are the subject of interpretation in the scientific literature. Also included is scientific description for research that is in the process of being reviewed by an institutional review board. Here, the institutional review board has as its federally mandated duty to protect the human subjects of research. The institutional review board reviews all human-subject scientific protocols and informed consent forms to initially approve them, to monitor ongoing research studies over time, and to conduct at least yearly reviews of ongoing studies. The research study and its informed consent form must be reviewed from both scientific and ethical standpoints, in terms of the scientific importance of the study related to the risks that human subjects may face during study participation and in terms of the adequacy of the informed consent form. It is the duty of the institutional review board to suggest modifications to the informed consent form such that it will be understood by the patient.

The informed consent form includes descriptions of

- the risks to human subjects;
- the study's purpose;
- the study's methods;
- alternatives to participating in research (that is, the options available in standard clinical care);
- risks that may accrue to the human subject during participation;
- the chance that the patient may not benefit at all from participation in the study;
- the laboratory and study testing required to be a participant in the study;
- the time line for the study;
- whom to contact regarding scientific or ethical questions during the course of the study (e.g., the study's principal investigator and the chairperson of the institutional review board);
- whether the patient will be compensated for injury or disability related to his or her participation in the research study.

Federal regulations allow the institutional review board to observe informed consent sessions between a principal investigator or his or her designee and individuals being recruited into the research study.

It is important to realize at the outset that the *understanding* of a patient regarding all aspects of his or her human-study participation is itself a subject of ongoing controversy and scrutiny. This controversy reaches perhaps its highest levels in the study of patients with mental health problems and psychiatric diseases.

For example, Paul Appelbaum has noted that "a recent study of decisional capacities in the inpatient treatment context demonstrated substantial impairments in approximately 52% of schizophrenic subjects and 24%

of depressed subjects, but only 12% of seriously medically ill subjects."[1] Thus, although a basic level of understanding needs to be achieved before a human subject should be enrolled in a scientific research protocol, the measurement of this *understanding* and its level at the time of enrollment in the research study is still controversial and highly debated.[2] Schizophrenia has been the test case in this arena because "schizophrenia has long been associated with abnormalities in information processing and attention mechanisms."[3]

Carl Salzman identifies articles and editorials by prominent psychiatric researchers documenting unacceptable ethical behavior in pharmaceutical research studies. Salzman notes that "these articles have focused primarily on challenge or discontinuation studies in schizophrenia," arguing that "while it is important to acknowledge that some psychiatric research studies may have posed risks to patients, risks that did not outweigh their benefits, it is equally important to emphasize that without research there will be no advancement in the treatment of serious mental disorders."[4]

Research studies that involve the use of placebo or that require patients to withdraw from effective medication (the "discontinuation studies" referred to by Salzman) continue to receive critical attention. For example, a withdrawal or discontinuation study may require a drug washout, wherein the drug or drugs that patients have taken for their care are eliminated through the processes of routine metabolism, so that the patient will be off all previous medication before a clinical trial is started. These types of studies are still actively controversial.[5] The issue is whether such studies should be done and, if so, under what circumstances. All groups need to be focused on answering the question of how to best protect the human subject during placebo studies or during studies requiring patients to be in a medication-free period. David Cohen has argued that "abrupt withdrawal increases the probability of recurrence of psychotic symptoms" and "withdrawal reactions appear to be especially common when atypical neuroleptics are abruptly withdrawn." Cohen also calls for "the clinical need for gradual, patient-centered drug withdrawal."[6]

As regards the necessity of basing medical decisions on a solid scientific foundation, consumers and patients must understand additional differences regarding what occurs in, for example, a *randomized clinical trial* and what is done in *standard clinical care*. Appelbaum has noted that "projects have been criticized for failing to distinguish clearly between those benefits and risks associated with research and those that would be obtained if the potential subject sought ordinary clinical care" and that "risks in general sometimes seem to be downplayed." "Both problems," he adds, "may be exacerbated when researchers are also in charge of subjects' clinical treatment."[7] Here, the individual doing the recruitment of a patient into a research trial is actually the physician who is in charge of the nonresearch clinical decisions related to the patient's overall mental health care.

Carol Svec notes additional problems patients may have in understanding the notion of *randomization*. Svec takes the example of a study comparing two treatments.

> [I]f Treatment A is tested only on men and Treatment B is tested only on women, the results cannot properly be compared because the groups used were so different. . . . The findings would tell you absolutely nothing about how Treatment B would work on men (or conversely, how Treatment A worked on women).

In a rigorous scientific study, Svec writes, "subjects are chosen from the target population as a whole and assigned to treatments randomly, the variability of the entire population is spread out evenly, and neither treatment gets an unfair advantage."[8]

2. Scientific Descriptions of Numbers and Numerical Estimates

The second type of scientific description we are interested in is the description of numbers and numerical estimates. Here we are most interested in the results of scientific studies. In the medical scientific literatures, these numbers take the form of statistics. On the issue of translation of scientific approach and numbers, Svec argues that

> statistics are used to express the degree of confidence researchers have in their results. In other words, statistics are tools that allow researchers to estimate how much of the difference they observe is just because people are different and how much is because they have figured out how to treat a disease. This can be mind-boggling for those of us who are comfortable with the certainty of "2 + 2 = 4."[9]

Svec also notes that "the way a study is designed, the kind of subjects included in the study, the way a treatment is administered, and uncontrolled outside influences (known in scientific circles as uncontrolled 'variables') can all alter the kinds of numbers that result from scientific research."[10]

While scientists and physicians may understand scientific communications based on numbers, it is clear that many patients do not want their physicians to use numbers in communications with them about the risks of invasive medical interventions. These patients prefer their physicians to describe risk in terms of verbal probability terms or expressions, like *rare, possible,* or *probable.* While medical professionals may assign similar numerical meanings to these verbal probability terms within particular contexts, it is clear that patients do not.[11]

This world we are interested in is a scientific world that is translated into the language of nonscientists in a variety of ways with a number of competing interests. At one extreme, there are short information messages with some selected scientific content that are provided by advertisers of scientific products. At the other extreme, there are an increasing number of patients wanting not only to receive information about risks, benefits, and alternatives but also wanting to share in decision making with their physicians. And while patients want to share in medical decision making, many still prefer to communicate using words like rare, possible, and probable, rather than communicating in terms of numbers like percentages.[12]

In terms of *scientific communication,* we will be interested in both written and spoken communications involving consumers, patients, physicians, media, scientists, product manufacturers, and advertisements from product manufacturers and from government agencies.

Federal regulations attempt to protect human subjects by placing an institutional review board in a primary oversight role over the principal investigator, his or her scientific protocol, and his or her informed consent form. In a strong sense, a key component of patient protection involves informing patients in a quite detailed way regarding all key aspects of the research endeavor in which they will participate. Thus, federal regulations protect patients by providing them with information about a scientific study involving human subjects, which the institutional review board has thoroughly reviewed from both its scientific and ethical standpoints. In addition, the institutional review board has made and suggested modifications to further clarify the research, its nature, risks, and any types of compensation available, among other issues. Thus, federal regulations protect in strong part by insisting that individuals being recruited into research studies are optimally informed about many key issues.

I state "many" key issues because many others are still not well clarified within federal regulations, particularly as regards patients with an impaired decisional capacity. But the *Code of Federal Regulations* attempts to protect human subjects of research through the disclosure of information of many types. Appelbaum sees even further roles for the institutional review boards in developing guidelines for the protection of decisionally impaired patients considering participation, "given that existing regulations grant IRBs substantial discretion to devise other means to achieve this goal."[13]

We will now contrast this notion of patient protection in clinical research with recent approaches to patient safety. The notion of safety is often juxtaposed with the notion of *disaster.* Eric Behrens argues:

> [S]ocieties in a pre-disaster state generally resist change, preferring instead the predictability of the familiar. Before change can occur, reformers must overcome the normal resistance to reform by creating an environment that is favorable to

their program. . . . [A] disaster threat is typically an "ill-structured problem," when even those who forecast the disaster cannot clearly define it. Consequently, individuals not only lack the concrete evidence necessary to dissuade them from their complacency, but also rely upon the poor definition of the problem as evidence that the forecasters' warnings are without substance. . . . [T]he disaster itself is a kind of conclusion, a final incident in the chain of evidence, and silences pre-disaster speculation with post-disaster direction.[14]

This is particularly true in the history of patient safety. Behrens notes that "individuals are particularly unlikely to understand the significance of warning signals when the threatening disaster is not of the frequently recurring sort, but is instead a rare and wholly unexpected calamity." Behrens points to the work of Dennis Wenger: "one disaster analyst notes that major disasters are amazing events because they depend upon a multitude of developments timed so perfectly that it is surprising that they ever occur at all." Behrens argues that "a disaster creates a climate uniquely conducive to social reform and legislation. In the immediate aftermath of major disasters, groups and individuals interested in reform are given unexpected opportunities to effect reforms that normally would take years to evolve."[15]

The disaster that prompted U.S. federal regulation in patient safety involved the drug thalidomide and the adverse side effect, phocomelia. However, Robert Veatch places the concerns about patient safety in light of the evaluative system of medicine, which allows physicians as learned intermediaries to make decisions regarding the use of drugs. Veatch argues as follows:

> While in unreconstructed modern medicine, safety and efficacy were thought to be scientific categories, in postmodern medicine, which is thought to be an evaluative enterprise, these are value terms. In fact, the very notion of whether an effect is a benefit or a harm is part of the evaluative process. . . . Likewise, to say an intervention is effective is a judgment based on how important the benefits are as well as exactly which benefits are on the agenda . . . What we mean when we say a drug is effective . . . [is] first, that the effect is a good one, and, second, that it produces the effect at some acceptable level of reliability.[16]

Veatch also notes that these decisions, involving the "use of drugs [in the care of patients can be] beyond the scope of what the drug has received federal approval for in terms of drug labeling." Veatch argues, "In the postmodern evaluative system of medicine, to say an intervention is *safe* is to say nothing more than that the possible outcomes are worth it based on the value system of the one making the judgment."[17]

More recently, the Institute of Medicine has juxtaposed the concept of *patient safety* and the *reduction of errors in medicine*. Kohn, Corrigan, and Donaldson argue that "in terms of lives lost, patient safety is as important an issue as worker safety. Every year, over 6,000 Americans die from workplace injuries. Medication errors alone, occurring either in or out of the hospital, are estimated to account for over 7,000 deaths annually." Furthermore, "the decentralized and fragmented nature of the health care delivery system (some would say 'non-system') also contributes to unsafe conditions for patients and serves as an impediment to efforts to improve safety." Kohn et al. define safety as "freedom from accidental injury," which they say "is the primary safety goal from the patient's perspective."[18]

The mission statement of the National Patient Safety Foundation includes the following:

> When a health care injury occurs, the patient and the family or representative are entitled to a prompt explanation of how the injury occurred and its short- and long-term effects. When an error contributed to the injury, the patient and the family or representative should receive a truthful and compassionate explanation about the error and the remedies available to the patient. They should be informed that the factors involved in the injury will be investigated so that steps can be taken to reduce the likelihood of similar injury to other patients. Health care professionals and institutions that accept this responsibility are acknowledging their ethical obligation to be forthcoming about health care injuries and errors.[19]

Yet patient safety can and should be approached prospectively, before injuries occur. This book focuses on the *prospective role* in which information, in particular medical scientific information, impacts patient and physician decision making. We discuss the range of sources of medical scientific information in contemporary society, from subsets of medical scientific information in advertisements (aimed at consumers and patients, as well as physicians and scientists) to the notion of shared decision making between patients and physicians.

We are in a stage now, following the 1980 consumer revolution in health care, where patient safety is again becoming a paramount interest in medicine. Yet patient safety is balanced against the *individual's right to assume risks*. The issue here is quite complex, because the assumption of risk by the individual needs to be based on consideration of scientific information about the danger in question. An individual can abrogate his or her right to receive scientific information, but then the competent adult must accept the possible adverse outcomes. This is why the communication of scientific information becomes the key focus in any form of medical decision making. In order for competent adults to understand the risks they are assuming, there must be optimal communication of scientific information about the competency of the physician performing the intervention, the risks and benefits of the procedure, the reasons for the timing of the intervention, and the reason why an intervention should be undertaken at all.

The interface of patient safety and the individual assumption of risk through the communication of scientific information becomes the focus of an ethics of scientific communication, which needs to address the following issues:

- the scientific content of the information;
- who is conveying the information;
- who is receiving the information;
- the authority of the information;
- the accuracy of the information;
- the "rules" used to select the particular subset of information that is communicated.

Obtaining optimal answers to this set of questions becomes the underlying theme of an ethics of scientific communication.

Veatch asks the following question of medical ethics: "Is the subject matter of medical ethics the personal moral requirements of the physician in relation to the patient, or does the concept include a social, political dimension that extends beyond personal ethics?"[20] I will take the perspective that the ethics of scientific communication has many different dimensions. The optimization of communication needs to be an object of continued study and improvement for all parties involved in the communication. This is particularly true of scientific communication where there may be a knowledge gap between, on one side, the physician and the scientist and, on the other side, the consumer and patient.

Historically, this issue of scientific communication has not taken the form of communication at all, but rather it has been discussed in terms of *disclosures*, particularly disclosures of risk information. For example, the National Research Council's Committee on Risk Perception argued in 1989 that "in the past, the term *risk communication* has commonly been thought of as consisting of one-way messages from experts to non-experts."[21]

The *flow of information* can be of two basic types:

- information flow from an expert source (e.g., an expert or expert panel) directly to a recipient (e.g., a consumer or patient);
- information flow from a product manufacturer, through a *learned intermediary* to a recipient.

For any flow of information to be truthfully conveyed or transmitted, that information flow must be understood *outside of product advertising* (e.g., as a system of communication that must exist between experts of various types, intermediaries of various types, and consumers and patients), as well as *within product advertising*. In each case, because of time constraints, the information must be short, and here is where the problem lies. Key to this framework is an understanding of the ethics of scientific communication and an examining of the multiple perspectives that influence scientific communication, especially in a short communication.

Notes

1. P. S. Appelbaum, "Rethinking the Conduct of Psychiatric Research," *Archives of General Psychiatry* 54, no. 2 (February 1997): 117–20. The study referred to is T. Grisso and P. S. Appelbaum, "The MacArthur Treatment Competence Study, III: Abilities of Patients to Consent to Psychiatric and Medical Treatments." *Law and Human Behavior* 19, no. 2 (April 1995): 149–74.

2. See, for example, R. R. Faden and T. L. Beauchamp, *A History and Theory of Informed Consent* (New York: Oxford University Press, 1986); E. J. Emanuel and F. G. Miller, "The Ethics of Placebo-Controlled Trials: A Middle Ground," *New England Journal of Medicine* 345, no. 12 (September 20, 2001): 915–19; and D. J. Mazur, *Shared Decision Making in the Patient–Physician Relationship: Challenges Facing Patients, Physicians, and Medical Institutions* (Tampa, Fla.: American College of Physician Executives, 2001).

3. E. Kraepelin, *Dementia Praecox and Paraphrenia*, trans. R. M. Barclay (1913; Edinburgh: E. & S. Livingston, 1919). See also E. Bleuler, *Dementia Praecox or the Group of Schizophrenias*, trans. J. Zinkin (1911; New York: International Universities Press, 1950); and V. Kumari, W. Soni, and T. Sharma, "Normalization of Information Processing Deficits in Schizophrenia with Clozapine," *American Journal of Psychiatry* 156, no. 7 (July 1999): 1046–51.

4. C. Salzman, "Not All Psychiatric Research Is Bad!" *American Journal of Psychiatry* 156, no. 7 (July 1999): 987–88.

5. S. Kumra et al., "Including Children and Adolescents with Schizophrenia in Medication-Free Research," *American Journal of Psychiatry* 156, no. 7 (July 1999): 1065–68.

6. D. Cohen, "Medication-Free Minors with Schizophrenia," *American Journal of Psychiatry* 158, no. 2 (February 2001): 324.

7. Appelbaum, "Rethinking the Conduct of Psychiatric Research," quotes from p. 117.

8. C. Svec, *After Any Diagnosis* (New York: Three Rivers Press, 2001), quotes from p. 246.

9. Svec, *After Any Diagnosis*, 244.

10. Svec, *After Any Diagnosis*, 245.

11. See A. Kong et al., "How Medical Professionals Evaluate Expressions of Probability," *New England Journal of Medicine* 315, no. 12 (September 18, 1986): 740–44; and

D. J. Mazur and D. H. Hickam, "Patient Interpretations of Terms Connoting Low Probabilities When Communicating About Surgical Risk," *Theoretical Surgery* 8 (1993): 143–45.

12. D. J. Mazur and D. H. Hickam, "Patient Preferences for Risk Disclosure and Role in Decision Making for Invasive Medical Procedures," *Journal of General Internal Medicine* 12, no. 2 (February 1997): 114–17.

13. Appelbaum, "Rethinking the Conduct of Psychiatric Research," 118.

14. E. G. Behrens, "The Triangle Shirt Company Fire of 1911: A Lesson in Legislative Manipulation," *Texas Law Review* 62 (1983): 361–87, quoted from pp. 373–74.

15. Behrens, "Triangle Shirt Company Fire." The work cited by Behrens is D. Wenger, "Community Response to Disaster: Functional and Structural Alterations," in *Disasters: Theory and Research*, ed. E. L. Quarantelli (London: Sage, 1978), 17–47. See also N. Pidgeon, K. Henwood, and B. Maguire, "Public Health Communication and the Social Amplification of Risks: Present Knowledge and Future Prospects," in *Risk Communication and Public Health*, ed. P. Bennett and K. Calman (New York: Oxford University Press, 2001), 65–77.

16. R. M. Veatch, *The Patient–Physician Relationship: The Patient as Partner, Part 2* (Bloomington: Indiana University Press, 1991), 273.

17. Veatch, *Patient–Physician Relationship, Part 2*, 273.

18. L. T. Kohn, J. M. Corrigan, and M. S. Donaldson, eds., *To Err Is Human: Building a Safer Health System* (Washington, D.C.: National Academy Press, 1999), 4.

19. Mission statement of the National Patient Safety Foundation, at www.npsf.org (accessed 3 July 2002).

20. Veatch, *Patient–Physician Relationship, Part 2*, 17–18.

21. National Research Council Committee on Risk Perception and Communication, *Improving Risk Communication* (Washington, D.C.: National Academy Press, 1989), 2.

4

The Ethics of Scientific Communication: Its Interdisciplinary Nature

THE ETHICS OF SCIENTIFIC COMMUNICATION is not solely an ethical topic. The issues involved intersect multiple disciplines, including science and the quality of scientific evidence, law and judicial decision making, philosophy and ethical decision making, the decision sciences, cognitive psychology, and all groups in relation to the Internet.

Scientific Information

We will use the term *scientific information* in a broad sense, as the evidence that is available based on good science. Why not the "best" science? Because the "best" science is not always available and may be costly to fund and acquire. In addition, the "best" science may be considered too risky by institutional review boards and human subjects. Thus, for our purposes, scientific information is data that result from scientific studies. Here it is important to recognize that

- The quality of the data that result from a study are highly dependent on the methods selected.
- Once scientific data are obtained, this data must be scrutinized.
- Once scrutinized, scientific data may be subject to multiple interpretations.

Finally, it must always be borne in mind that there are many gradations of scientific evidence and many ways to judge the import of scientific evidence in terms of its quality and applicability to desired ends.

Science also includes within its meaning the problems of uncertainty inherent in medical decision making and in decision making underlying research on human subjects.

Perspective 1. Scientific Views: Clinical Science vs. Clinical Practice

Science is thought of as an area of precision—precision of description and precision of measurement. Medical science is often seen as an area of imprecise descriptions and imprecise measurement in certain of its domains. While a heart catheterization laboratory may provide precise anatomical descriptions of the location of lesions on a patient's coronary arteries, the setting of congestive heart failure trials may provide less precise information. Why? Because while the cardiac catheterization procedure is under the control of the physician–scientist, the clinical drug trial showing potential impacts of new drugs on the survival of heart failure patients is more responsive to the preferences of the patients participating in the study. Nonetheless, the patient has to abide by the restrictions imposed by the protocol, for example, restrictions on the consumption of salt (sodium chloride) and salty foods.

At present, there is an active controversy regarding two areas of medical professionalism. Is the best clinician the best interpreter of the existing scientific medical evidence, or is the best clinician the one with the most clinical experience in patient care? On the side of the physician as interpreter of evidence, Liz Trinder argues that

> evidence-based practice . . . throws down the gauntlet to other traditional authorities, the leaders of professions steeped in experience and authority but not necessarily in the best evidence. In the emphasis on self-learning, and the belief that anyone can learn the skills of evidence-based practice, it is potentially therefore a radically democratizing strategy where the most junior members of the profession can be as skilled in identifying the evidence as the most respected. Evidence rather than experience becomes privileged.[1]

Yet, there is a paucity of high quality scientific evidence on many clinical topics, so that there must be a sharing of authority between scientific evidence and clinical experience, where experts are needed in both domains of inquiry.

We are interested in scientific information in the domain of the clinical researcher and the domain of the clinical practitioner. The clinical researcher, in the main, studies a medical condition and the impact of treatment in a highly select population of patients. It is important to consider this population of study as highly select because precise selection of patients is an important part of the conduct of the science of medicine. For example, in studying new drug therapy, when an adverse outcome occurs, the principal investigator must be able to determine whether that adverse outcome was "caused" by the new drug. What are the other options in a clinical trial involving a comparison of a new drug or a drug used in standard therapy? The adverse outcome could be caused by some other aspect of the patient's medical condition, or by some other medical conditions a patient may have. Or, the patient could in fact have sustained the adverse outcomes because yet-to-be-understood causal conditions exist between the drug and the patient's medical condition. Thus, patients with multiple medical conditions may be excluded from study participation so that the physician–scientist can be confident that observed benefits or adverse outcomes are attributable to the drug being studied.

The domain of clinical practice involves applying what has been learned in the clinical sciences to the care of individual patients. When a drug is approved by the FDA and marketed, there is much that is still not known about it. Why? Well, we have already reasoned that patients with several medical conditions or patients taking several drugs for their medical condition may be excluded from the clinical trial in order for the principal investigator to "best decide" whether a benefit accrual or an adverse outcome was "caused" by the new drug. Therefore, it is not until after a drug is approved and marketed that the patients with multiple medical conditions or taking multiple drugs get the new drug for the first time. Yet this drug has not necessarily been studied in complex patients in any of its trial phases.

J. A. Muir Grey notes that studies have demonstrated much greater variations in clinical practice and the rates of intervention in patients than can be explained by variations in need of different populations. Muir Grey argues that "clinical practice [is] not as cut-and-dried as the image of modern scientific medicine has suggested."[2] Here, *evidence-based medicine*, the conduct of medicine on the basis of the best scientific evidence available, has taken a hard look at whether and how scientific evidence is applied to clinical care. Yet, it must be remembered that there is also a science underlying the clinical experience of physicians. And this science of clinical experience also needs further development and refinement.

Perspective 2. Judicial Views

The concept of a physician needing to obtain his or her patient's consent to a medical intervention can be traced back to the late 1700s in Great Britain.[3] In these early cases the question was, "Did the doctor get the competent adult patient's consent before proceeding to intervene in the patient's behalf?" There was much less emphasis on what information physicians actually used to secure their patient's consent. The term *informed consent* entered the judicial lexicon in 1957 in a California appellate court decision.[4] The basic nature of the patient–physician relationship is one in which the physician is considered as a *fiduciary*, or someone acting in the patient's best interests. But while physicians understand medicine, courts have held that patients have the right to decide what is to be done to their own bodies. In Judge Benjamin Cardozo's written court opinion of 1914, the judicial concept of consent was firmly grounded on the principle of a patient's right to *self-determination*.[5]

In eras when the patient–physician relationship was defined in terms of the law and judicial decision making, the courts argued that there should be in the mind of the physician caring for a patient a clear distinction between the *competent adult patient* and the *patient who is incompetent*. However, this distinction is often very difficult to make when, for example, the patient has a medical or psychiatric illness, or is being evaluated in an emergency room. Research studies have suggested that patients with a medical illness may function at the level of a ten-year-old, as measured by Piaget's tasks.[6]

Between 1767 and 1972, there was only one standard of obtaining consent, the professional standard. That is, a physician was judged by the standards of peers in his or her area of medicine. This professional standard remains the only standard in Great Britain. It is held by a slight majority of states in the United States and was the standard in Canada until 1980.[7] The second major standard of informed consent, the *reasonable person standard*, was created by Judge Spottswood Robinson in his 1972 federal decision in the District of Columbia, *Canterbury v. Spence*.[8] The reasonable person standard is held by the Supreme Court of Canada. Great Britain still maintains only the professional standard.

We have so far discussed the obligations of a physician to the *competent adult patient*. When the patient develops a psychiatric illness, the question is often whether the patient has capacity to make a decision, and if so, whether the patient is competent to make a decision. This distinction between *capacity* and *competency* is key. *Capacity* refers to the patient's ability to make a decision on his or her own behalf, and *competency* is sometimes considered to refer to the quality of the decision the patient is making. As Tom Beauchamp and James Childress note, "several commentators distinguish judgments of capacity from judgments of competence on the grounds that health professionals assess capacity and incapacity, whereas courts determine competence and incompetence."[9] But Beauchamp and Childress agree with Thomas Grisso and Paul Appelbaum that "this distinction breaks down in practice: 'When clinicians determine that a patient lacks decision-making capacity, the practical consequences may be the same as those attending a legal determination of incompetence.'"[10] From a psychiatric perspective, Grisso and Appelbaum hold that the purpose of examining competence and the type of information to be explored with the patient involves more than a discussion of risks. They argue that such a discussion must "examine the degree to which patients are able to go beyond the ordinary medical and psychiatric benefits/risks that have been described, to generate their own inferences about specific, practical differences that various outcomes might have in their lives."[11] There are many layers to the notion of competence. First, it must be determined prior to the informed consent session. Second, it is involved in the informed consent session itself.

Beauchamp and Childress recognize that "the competence to decide is . . . relative to the particular decision to be made. . . . It depends not only on a person's abilities but also on how that person's abilities match the particular decision-making task he or she confronts."[12] Grisso and Appelbaum focus on the distinction between "task" and "demand": "Competence is not simply dependent on a person's abilities, but on the *match or mismatch between the patient's abilities and the decision making demands of the situation that the patient faces*" (italics in original). For Grisso and Appelbaum, this view of competence has important implications:

- There is no absolute level of ability that defines competence or incompetence.
- The degree of ability that is necessary in order to be considered competent will depend in part on how much is demanded.
- How much is demanded in a particular situation will vary from case to case with no single criterion applying across all cases.

Grisso and Appelbaum state that "two patients may have equal degrees of decision-making ability, yet one may be considered competent and the other incompetent, if the decisions that they face are very different from their demands."[13]

From a philosophical perspective, B. C. White considers competence in informed consent to include "eight distinct but functionally inseparable capacities: . . . to receive, recognize, and remember relevant information; to relate to oneself, reason about, and rank alternatives; to resolve situations; and to resign oneself to those resolutions."[14] Potential problems here involve the notions of resolving situations and resigning oneself to those resolutions. Since there are many uncertainties involved in an informed consent session in clinical care and clinical research, and many economic factors involved, many aspects of the decision may be very difficult to resolve and resign oneself to. The fight to define competence in informed consent and what constitutes an optimally informed consent has continued to evolve precisely because individuals engaged in the task both as patients and as commentators have not resigned themselves.

The establishment of a patient's competence has important consequences for that patient. According to Judge Robinson in *Canterbury v. Spence*, in the absence of emergency, of the patient being unconscious, of permanent cognitively disabling problems, and of the threat that risk disclosure itself poses a detriment to a particular patient (the "therapeutic exception"), the competent adult patient has the right of self-decision and may ground that decision on "any reason that appeals to him or her."[15]

Gert, Culver, and Clouser define competence as "the ability to make a rational decision," arguing that

> a person is competent to make a rational decision if the following is true: he or she does not have a cognitive disability preventing him or her from understanding and appreciating the relevant information or coordinating the information with his or her own stable values, and he or she does not have a mental malady involving a volitional disability that interferes with his or her ability to make a rational decision. If none of these disabilities, including a relevant mental malady, is present, he or she is competent to make a rational decision even if he or she is making an irrational decision.

Gert, Culver, and Clouser believe that "the ability to make a rational decision of a certain kind is what people have always had in mind when they accorded 'competence' the primacy it has in the consent process."[16]

The term *rational decisional capacity* can be used to describe the approach a rational decision maker may take to prioritizing and weighing information in attempting to choose, for example, among a set of alternative treatment options.[17] Adrian Edwards and Glyn Elwyn extend this use of prioritizing information beyond a purely medical sphere to the interface of medicine and economics, noting that "decision making by individuals has been widely discussed in economic contexts where the dominant theory is that of *rational choice* engaging in a free market environment."[18]

Rational decision making is dependent on quality information. When talking about rational decision making, patients may focus on their increased access to medical information on the World Wide Web. Yet Muir Gray argues that increasing access has nothing to do with the quality of the information:

> As there is no mechanism to control the content of material put on the Web, the information that patients are able to find can vary widely in quality. It is important, therefore, that health professionals contribute to providing good-quality information on the Web, i.e., information that is easy to find, easy to read, and free from bias.[19]

Physicians themselves may often have little high-quality, evidence-based information to rely on in helping determine the risks and benefits of the choices that their patients face. For example, research studies that have been done on human participants may not be usable. Halpern, Karlawish, and Berlin observe that "more than 20 years have passed since investigators first described the ethical problems of conducting randomized controlled trials (RCTs) with insufficient statistical power. Because such studies may not adequately test the underlying hypothesis, they have been considered 'scientifically useless.'"[20] And much of the information that patients and physicians would like from research studies is not available because the key research studies, such as head-to-head comparisons of medical products (devices and prescribed drugs), have not been done. The information is just not available for use in clinical care.

Carl Elliott continues to force us back into the realities of decision making and the impact of mood on competence.

> Most accounts of competence focus on intellectual capacity and abilities to reason, and depression is primarily a disorder of mood. According to conventional thinking, depression is primarily about despair, guilt, and a loss of motivation, while competence is about the ability to reason, to deliberate, to compare, and to evaluate.

Elliott argues that depression "can impair a person's ability to evaluate risks and benefits. To put the matter simply, if a person is depressed, he or she may be *aware* that a protocol carries risks, but simply not *care* about those risks."

> Sometimes depression is not even considered a warning sign that a person's competence to consent needs to be evaluated, largely because it is not thought to be the type of disorder that would ordinarily interfere with competence. It is the rare depressed patient who is psychotic, and while depression may often interfere with a person's memory and concentration, often this interference is not severe enough to raise any warning flags.[21]

Elliot is telling us that these warnings flags have to be raised.

It is important that our future research be sufficiently powered and directed at comparing medical products and treatments head-to-head, so that we have high-quality information on which to base our decisions as a society and as individuals with different health states and different values. We all—citizens, patients, physicians, and governments—are in this together, deciding what is best for the health and well-being of our population. But it is important to recognize that not only do we face problems at the extreme of rational decision making and challenges in understanding what constitutes optimal decision making, but we face these challenges in the more commonplace realm of depression and its impact on competence to consent.

Capacity is the ability to "understand, appreciate, reason."[22] The 1983 U.S. President's Commission held the view that "any determination of the capacity to decide on [or consent to or reject] a course of treatment must relate to the individual abilities of a patient, the requirements of the task at hand, and the consequences likely to flow from the decision." For this commission, "decision making capacity requires, to greater or lesser degree: (1) possession of a set of values and goals; (2) the ability to communicate and to understand information; and (3) the ability to reason and to deliberate about one's choices."[23] According to Elliott, "as conceptualized within this framework, a potential research subject takes in the relevant information, weighs it according to his or her goals and values, and then reasons to an informed decision."[24] However, Elliott argues that "if competence to consent to research is defined simply as the ability to make a decision to enroll in a research protocol, we face the familiar problem of what *counts* as that ability. . . ." Is a person to be judged incompetent, Elliott asks,

- by virtue of making a poor decision,
- by virtue of making an irrational decision, or
- by virtue of coming to that decision in an unsystematic, illogical, or erratic way?

Elliott goes on to elaborate on the complexity of this *counting* task:

> Since even competent people are sometimes stubborn, obtuse, or unreasonable, we need an account of competence that explains why we sometimes believe that a person can be both competent and make bad, irrational, or even unreasonable decisions. . . . What we really want to know when we ask if a patient is competent is whether that individual is able to make a decision *for which he or she can be considered accountable*. . . . [I]f a person is making a decision that will affect his life in momentous ways, we will naturally be concerned that he makes a sound decision. But because we recognize that a person generally has the right to make even unsound decisions, a judgment about competence ensures that whatever decision a person makes, it is truly his or her decision: a decision for which he or she can finally be held accountable.[25]

Capacity can be enhanced by providing information in an educational context and by helping to clarify the consequences of participation by including significant others or advocates in the decision-making process. William Carpenter Jr. notes that "adequate decision-making capacity for providing informed consent to research participation can be assessed and documented." But Carpenter also has a series of questions:

> [H]ow well is this being done in all the various settings where research is conducted? What constitutes adequate capacity, and how is this to be determined and documented? Who should participate in informed consent, and how should

research be conducted if the person is judged to be too impaired for competent consent? How should these procedures be reviewed, and which stakeholders should participate in the review?[26]

It is important to understand that there is active controversy regarding what is meant by the terms *capacity* and *competency*. This controversy is found in both the clinical care sector, in judging whether a patient has the capacity and competency to consent to prescription medications, and in clinical research, particularly in the area of judging whether a patient has both the capacity and competency to enroll as a human subject in scientific research studies.

While the judicial concept of consent was firmly grounded by Judge Cardozo in 1914, in the principle of a patient's right to *self-determination*,[27] Judge Robinson argued that in the notion of "information + consent"—that is, *informed consent*—the patient had the right to information disclosed by the physician because of the patient's right to *self-decision*.[28] For Judge Robinson, the competent adult patient has a right to make a decision on whatever grounds, rational or irrational, the patient sees fit.

Perspective 3. Ethical Views

Ethical views have focused on developing the concepts of *autonomy* and *benevolence* in the areas of the patient–physician relationship. While the courts have focused on *information disclosure* by the physician, ethicists like Ruth Faden and Tom Beauchamp have focused on the need of the patient to *understand* the disclosed information.[29] But information disclosure has remained the dominant emphasis of the courts, and patient understanding of that information a dominant ethical standard. While Faden and Beauchamp have argued for the need for patients to have *substantial understanding* in relation to informed consent, what is still needed is an understanding of the impacts of information in the patient–physician dialogue.

Some questions remain after understanding the judicial and ethical perspectives on information and decision making. What are the cognitive problems that exist in communicating scientific information in the patient–physician dialogue or the principal investigator–human subject informed consent session? How are these cognitive challenges regarding information and choice to be addressed? Proposed solutions have come from two groups: decision scientists and cognitive psychologists. The decision scientists point to the importance of systematic approaches to information and the structure of the medical decision. Cognitive psychologists point to the problems that exist in the presentation of information to patients and research subjects and in successfully processing information for use in choice and treatment decisions.

Perspective 4. Views of the Decision Sciences

Decision scientists have offered another perspective that has carried over to the issue of informed consent. That is, what is needed to let a patient make the best decision? Here systematic approaches of decision analysis have a potential role in clarifying some issues of informed consent.[30] The rational decision maker of the decision scientists is usually approached through a systematic methodology of eliciting a patient's preferences for a set of outcomes, including standard gambles and time trade-offs.[31] While the perspective of the rational decision maker may be difficult to pursue by the bedside of the hospitalized patient, similar methodologies have been incorporated in decision making "tools" and "decision support systems" for use by the patients in supervised and unsupervised settings.[32]

Perspective 5. Cognitive Psychological Views

Psychologists, especially cognitive psychologists, have focused on three key areas related to the ethics of scientific communication: the *shortcuts* human beings take when making choices, the impact of the way information is *framed* or *presented* on patient choice, and the challenges *human memory* provides to the patient trying to listen and understand the scientific information disclosed to him or her by a physician or principal investigator.

Cognitive psychologists have found that respondents to surveys make systematic errors in choice situations.[33] Kahneman, Slovic, and Tversky argue that humans do not follow the rational principles of the decision sciences in the choices that they make.[34] Rather, human decision making is fraught with shortcuts that yield suboptimal decisions. In addition, cognitive psychologists have examined how the way information is presented to humans influences human choice.

Tversky and Kahneman have done work on the *availability bias*. Here decision makers assess the frequency of a class or the probability of an event by "the ease within which instances or occurrences can be brought to mind."[35] Scott Plous emphasizes the point that "some events are more available than others *not* because they tend to occur frequently or with high probability, but because they are inherently easier to think about, because they have taken place recently, because they are highly emotional, and so forth."[36]

Cognitive psychologists have also recognized the impact of *framing effects* or *presentation effects* that influence the way patients make choices. When information is framed or presented in a certain way, this frame of presentation can unfairly influence the way the information is treated by the patient. Yet there is more than simply a framing problem in communication about scientific information. Humans still are plagued by shortcuts in decision making that may yield choices that do not reflect actual patient preferences. It must be recognized that much of this research has been done in relation to hypothetical treatment decisions, although more research is being conducted in actual situations where patients are making decisions.

Also important is the understanding of human memory, especially patients' recall of the information they have received about risks and benefits of proposed medical interventions. Valerie Reyna and Allan Hamilton argue that "memory research over the past decade has shown that people encode separate verbatim memories of quantitative information, such as probabilities as well as gist (or approximate) memories, and that decisions are based on the gist memories rather than verbatim ones." The authors find that "verbatim memories represent the facts and details on information as it was presented, but they fade rapidly; gist memories, which reflect understanding and interpretation, endure."[37]

Finally, even if framing and decision shortcuts are accounted for, there are issues of *how stable* patient preferences are over time. In particular, how does new information about risk and benefit influence patient preferences that seem to have been stable over time, for example, to a physician caring for a patient over many years?[38]

Perspective 6. Federal Regulation of Research Involving Human Subjects: The Belmont Report

The Belmont Report forms the philosophical basis of the federal regulation of research in the United States. The report focuses on research on human subjects and can be seen as developing another standard, the *reasonable volunteer standard*, in clinical research. It argues that the nature of information should be such that persons, knowing that a procedure is neither necessary for their care nor fully understood scientifically, can decide whether they wish to participate in the furthering of scientific knowledge. Even when some direct benefit to the patient is anticipated from the research, the Belmont Report holds out one of the broadest disclosure standards. While the standards of informed consent in clinical care are circumscribed by the professional and reasonable person standards plus possible exclusions related to commonly known risks, the Belmont Report argues that study participants need to be informed of the "range of risk" associated with their participation in the research.[39]

Perspective 7. Federal Regulation of Prescription Drugs and Free-Market Communication

The final perspective relates to federal regulations that have a direct impact on what information, particularly scientific information, is allowed to be presented to the public about prescription drugs in advertisements in the televised and radio media. Federal regulations pertaining to direct-to-consumer advertising allow pharmaceutical companies to develop short information messages that contain selective types of information related to a drug. There are federally regulated aspects of medical products used in what I call *free market communication*, where pieces of scientific information are subsumed within an overall advertising message.

Thus, as we begin to view the patient–physician dialogue, more has to be accounted for than the medical information disclosed by the physician. The Internet has added to the availability of information but not to the scientific clarity of that information. The questions related to information include:

- What is *full disclosure* of information?
- Who selects the information presented and what are the selection rules?
- Who decides how information is going to be presented?

Questions related to decision making include:

- How are patients' reactions to risk information to be accounted for in information presentation and decision making?
- When and where is information to be presented and decision making to take place?
- Who is going to present information, and what are the presenter's conflicts of interests?
- Once information is presented, how are patients to be protected from the shortcuts in decision making that plague all human choice?

Notes

1. L. Trinder, introduction to *Evidence-Based Practice: A Critical Appraisal*, ed. Trinder and S. Reynolds (Malden, Mass.: Blackwell Science, 2000), 11–12.

2. J. A. Muir Gray, "Evidence-Based Public Health," in *Evidence-Based Practice: A Critical Appraisal*, ed. Trinder and S. Reynolds (Malden, Mass.: Blackwell Science, 2000), 89–100, quoted from p. 92.

3. *Slater v. Baker and Stapleton*, 95 Eng. Rep. 860, 2 Wils. KB 359 (1767).

4. *Salgo v. Leland Stanford Junior University Board of Trustees*, 154 Cal. App. 2d 560, 317 P.2d 170 (1957).

5. *Schloendorff v. Society of New York Hospitals*, 211 NY 125; 105 NE 92 (1914).

6. E. J. Cassell, A. C. Leon, and S. G. Kaufman, "Preliminary Evidence of Impaired Thinking in Sick Patients," *Annals of Internal Medicine* 134, no. 12 (June 19, 2001): 1120–23.

7. *Hopp v. Lepp* [1980], 112 D.L.R. (3d) 67 (SCC); *Reibl v. Hughes* [1980], 114 DLR (3d) 1 (SCC); [1980] 2 SCR 880.

8. *Canterbury v. Spence,* 464 F 2d 772 (1972).

9. T. L. Beauchamp and J. F. Childress, *Principles of Biomedical Ethics,* 5th ed. (New York: Oxford University Press, 2001), 69.

10. T. Grisso and P. S. Appelbaum, *Assessing Competence to Consent to Treatment: A Guide for Physicians and Other Health Care Professionals* (New York: Oxford University Press, 1998), 11.

11. Grisso and Appelbaum, *Assessing Competence to Consent,* 118.

12. Beauchamp and Childress, *Principles of Biomedical Ethics.*

13. Grisso and Appelbaum, *Assessing Competence to Consent,* 23.

14. B. C. White, *Competence to Consent* (Washington, D.C.: Georgetown University Press, 1994), 144.

15. *Canterbury v. Spence,* 798–99.

16. B. Gert, C. M. Culver, and K. D. Clouser, *Bioethics: A Return to Fundamentals* (New York: Oxford University Press, 1997), 145–46.

17. See D. J. Mazur, *Shared Decision Making in the Patient–Physician Relationship: Challenges Facing Patients, Physicians, and Medical Institutions* (Tampa, Fla.: American College of Physician Executives, 2001).

18. A. Edwards and G. Elwyn, eds., *Evidence-Based Patient Choice: Inevitable or Impossible?* (New York: Oxford University Press, 2001), 3.

19. J. A. Muir Gray, *Evidence-Based Healthcare: How to Make Health Policy and Management Decisions,* 2d ed. (New York: Churchill Livingstone, 2001), 345.

20. S. D. Halpern, J. H. T. Karlawish, and J. A. Berlin, "The Continuing Unethical Conduct of Underpowered Clinical Trials," *Journal of the American Medical Association* 288, no. 3 (July 17, 2002), 358–62, quoted from p. 358.

21. C. Elliott, "Caring About Risks: Are Severely Depressed Patients Competent to Consent to Research?" *Archives of General Psychiatry* 54, no. 2 (February 1997): 113–16, quoted from p. 113.

22. G. Smukler, "Double Standard on Capacity and Consent?" *American Journal of Psychiatry* 158, no. 1 (January 2001): 148–49.

23. U.S. President's Commission for the Study of Ethical Problems in Medicine and Biomedical and Behavioral Research, *Making Health Care Decisions: A Report on the Ethical and Legal Implications of Informed Consent in the Patient–Practitioner Relationship,* vol. 1 (Washington, D.C.: The Commission, 1982), 57.

24. Elliott, "Caring About Risks," 114.

25. Elliott, "Caring About Risks," 114.

26. W. T. Carpenter Jr., "The Challenge to Psychiatry as Society's Agent for Mental Illness Treatment and Research," *American Journal of Psychiatry* 156, no. 9 (September 1999): 1307–10, quoted from p. 1309.

27. *Schloendorff v. Society of New York Hospitals.*

28. *Canterbury v. Spence.*

29. R. R. Faden and T. L. Beauchamp, *A History and Theory of Informed Consent* (New York: Oxford University Press, 1986).

30. See D. J. Mazur, "Informed Consent: Court Viewpoints and Medical Decision Making," *Medical Decision Making* 6, no. 4 (October/December, 1986): 224–30; J. Dowie and A. S. Elstein, eds., *Professional Judgment: A Reader in Clinical Decision Making* (New York: Cambridge University Press, 1988); and P. A. Ubel and G. Loewenstein, "The Role of Decision Analysis in Informed Consent: Choosing Between Intuition and Systematicity," *Social Science and Medicine* 44, no. 5 (March 1997): 647–56.

31. See H. C. Sox Jr., et al., *Decision Making* (Boston: Butterworth-Heinemann Medical, 1988); and J. S. Hammond, R. L. Keeney, and H. Raiffa, *Smart Choices: A Practical Guide to Making Better Decisions* (Boston: Harvard Business School Press, 1999).

32. D. J. Mazur, *Shared Decision Making in the Patient–Physician Relationship: Challenges Facing Patients, Physicians, and Medical Institutions* (Tampa, Fla.: American College of Physician Executives, 2001).

33. See B. Fischhoff et al., "How Safe Is Safe Enough? A Psychometric Study of Attitudes Towards Technological Risks and Benefits," *Policy Sciences* 9 (1978): 127–52; and P. Slovic, B. Fischhoff, and S. Lichtenstein, "Facts and Fears: Understanding Perceived Risk," in *Society Risk Assessment: How Safe Is Safe Enough?*, ed. R. C. Schwing and W. A. Albers (New York: Plenum, 1980).

34. D. Kahneman, P. Slovic, and A. Tversky, *Judgment Under Uncertainty: Heuristics and Biases* (New York: Cambridge University Press, 1982).

35. A. Tversky and D. Kahneman, "Judgment Under Uncertainty: Heuristics and Biases," *Science* 185 (1974): 1124–30.

36. S. Plous, *The Psychology of Judgment and Decision Making* (New York: McGraw-Hill, 1993), 121.

37. V. F. Reyna and A. J. Hamilton, "The Importance of Memory in Informed Consent for Surgical Risk," *Medical Decision Making* 21, no. 2 (March/April 2001): 152–55, quoted from p. 152.

38. See S. J. Jansen et al., "Patients' Preferences for Adjuvant Chemotherapy in Early-Stage Breast Cancer: Is Treatment Worthwhile?" *British Journal of Cancer* 84, no. 12 (June 15, 2001): 1577–85.

39. National Commission for the Protection of Human Subjects of Biomedical and Behavioral Research, *The Belmont Report: Ethical Guidelines for the Protection of Human Subjects of Research*, DHEW Publication, o.s., 78-0012 (Washington, D.C.: The Commission, 1978), 11. Accessible online at http://ohsr.od.nih.gov/mpa/belmont.php3.

5

Perspectives on Information and the Scientist, the Social Scientist, and the Philosopher

THE POTENTIAL IMPACT OF INFORMATION on choice circumstances and the relation of how the format or presentation of information influences choice had its origins in the landmark research of psychologists Kahneman, Slovic, and Tversky,[1] among others. More recently, there has been an increased interest in the impact of information and its presentation to patients in choice and decision-making circumstances with medical and surgical themes.

But social scientists are also interested in the patient–physician dialogue.[2] The interests of social scientists have spanned many domains, communication in the patient–physician relationship to the more recent areas of shared decision making in the patient–physician relationship.[3] Future research will no doubt proceed into the areas of the new conversation in medicine, where patients and physicians discuss not only medical therapies, but physician opinions on the use of nonmedical therapies, most specifically, complementary and alternative medicine (CAM therapies) in relationship to cancer, chronic pain, and specific areas of medicine and surgery (e.g., otolaryngology).[4] In addition, philosophers have been interested in conversations between patients and physicians,[5] not only in clinical care but also in clinical research.

What is the "new medical conversation" in clinical care and clinical research, and why is it coming about? In my opinion, this new development is related to one thing and one thing only: the increased access of individuals to *information*. And in the specific case of clinical care and clinical research, this means the increased access of patients to medical information. In the new medical conversation, new information is brought into the patient–physician conversation by patients.

Where is this "new medical information" coming from? In the main, there is one major source: the *advertised message*. But there are new twists that have been added to this traditional source of information. In the last few years, there has been an intensification of medical advertisements (predominantly for drugs and devices) encouraging the viewer or listener to "be sure to ask your doctor if this drug (or product) is for you." This type of promotional advertising was permitted with the passing of the FDA Modernization Act of 1997, which allowed the FDA to relax previous restrictions on the content of direct-to-consumer broadcast advertisements for prescription drugs and devices. These new rules have made it easier for product manufacturers to use "indication" advertisements, which mention both the name of the drug and its indication.[6]

For example, an advertisement for a drug to stimulate red blood cells in cancer patients may indicate that the drug may counteract weakness that a cancer patient may be experiencing during treatment. The individual seeing the advertisement may recall only this indication, without understanding that the drug has been approved for use only within a specific range of medical conditions.

Medical advertisements in magazines often include, in difficult-to-read fine print, a "brief summary" of side effects and adverse outcomes. This brief summary was derived from the FDA-approved label of the drug. Such summaries are not part of radio and television advertisements, although a brief mention of some of the common side effects and severe outcomes may be included.

The new FDA guidelines require that a broadcast advertisement not mislead the public, that it be truthful, and that it include within the advertisement "a major statement prominently disclosing all the major risks associated with the drug." In addition, the advertisements must "provide a mechanism to ensure that consumers can easily obtain full product labeling."[7] However, it is unclear how the FDA defines "not misleading," "truthful," and "major risks," making it difficult, if not impossible, to enforce these standards. To meet the resource requirement, a broadcast advertisement may offer three or four ways for consumers to access more complete information about the drug. But how can the goal of allowing consumer access to more complete information be accomplished within the time span of a short information message when there may be major differences of opinion on what to include within the scope of the term *major risk*? Does the term mean "major risk in a population of patients"? If so, what is the population of patients to be considered? Is this population the original study population of human subjects? Or is it the population of all patients in the United States who have used the drug since it has been on the market?

Phyllis Maguire, writing for the American College of Physicians and American Society of Internal Medicine's (ACP-ASIM) *Observer*, notes that in August 1997, the FDA began allowing TV and radio commercials to tout drugs' benefits without a lengthy summary of potential side effects and contraindications. Instead, broadcast advertising is required to mention only a drug's major risks and provide a Web address and toll-free phone number for consumers to go get more information.[8]

The Center for Medical Consumers' publication, *HealthFacts*, makes the following points:

[T]he FDA does not have enough resources to monitor the proliferation of ads likely to result from the new guidelines. The initial experience with TV ads demonstrates that industry cannot be relied upon to follow the guidelines. For example, the first Claritin TV indication ads gave only an Internet resource; other required content was in hard-to-read white type on a white background.

Furthermore, most advertising is for new drugs whose risks are not as well understood as those of older drugs. And the new guidelines require only "major" risks be mentioned, which means that consumers are likely to remain unaware of uncommon, but potentially serious adverse reactions that did not show up in the studies submitted for FDA approval. . . .

In the face of increasingly sophisticated and aggressive direct-to-consumer advertising, it is more important than ever for consumers to seek all available information from reliable resources. Most drug advertising takes benefit and risk information out of any scientific context and presents it so that the consumer "take-away" message plays up benefits. Brief summaries do not describe the findings of the effectiveness studies submitted by a manufacturer to the FDA for approval. As a result, [brief summaries] do not help consumers who want a more objective overview of a drug's proven benefits.[9]

In the new medical conversation between the patient and physician, even with direct-to-consumer broadcast advertising in full force, patients are still dependent on the physician fitting their particular case into an up-to-date understanding of the peer-reviewed medical literature.

We are beginning to see what had been a brief summary of a drug's potential downsides now eliminated in the radio and television broadcast message, so that the physician again becomes the informational intermediary between the patient and the drug or device. The purchase of a prescription drug is unlike the purchase of a power saw or power nailer. The manufacturer of those tools has a duty to warn the consumer and "end user" directly about hazards associated with their use. The manufacturer must also provide precautionary information about how to prevent foreseeable accidents. Some simple precautions include wearing safety glasses and plugging into three-hole, grounded electrical outlets. However, with prescription drugs, it is the physician who interprets the scientific/medical information supplied by the manufacturer in the labeling. As a learned intermediary, the physician must weigh the potential benefits of drug therapy against the possible risks of toxic reactions and injury, and make his or her best recommendation about appropriate medication. Because specialized learning is required to select the proper medication, the pharmaceutical manufacturer has a duty to warn only the physician directly, not the patient.

Yet specific drugs do not fall under this "learned intermediary" doctrine. For example, exceptions exist for drugs like oral contraceptives, IUDs, conjugated estrogens, and some other drugs like Accutane for cystic acne, when the patient is generally healthy and has significant input into the selection of treatment. With these latter types of drugs, the FDA requires the manufacturer to warn the patients, as well as the prescribers, directly.

Once provided with the safety and efficacy data in the labeling, the physician or prescriber then assumes the duty of warning the patient about the potential hazards of therapy during the informed consent process with the patient.

Whether one is a proponent or a detractor of direct-to-consumer advertising, broadcast advertising is only one of the sources of information resulting in consultations. For example, an increasing number of patients have access to websites of medical universities and medical centers; medical, surgical, and psychiatric associations; and federal agencies (e.g., the National Institute of Health and the National Institute of Mental Health) that identify and recommend published articles as potential sources of information for patients as consumers. Despite the research of social scientists on key questions facing the doctor and patient, in reality little is known about the origins, uses, or understanding of information the patient obtains before the patient–physician encounter begins.

Medical Information and Philosophers: Material Information and Perspectives on the Ethics of Informed Consent

Ruth Faden and Tom Beauchamp discuss the notion of informed consent primarily in terms of whether the patient understands disclosed information, and understands when he or she is authorizing a physician to begin a medical intervention by consenting. The authors note the problems regarding what it means for a patient to sign an informed consent form. They point out that physicians view the concept of informed consent more in terms of "informing patients about conditions and treatments" than as "patients authorizing interventions."[10]

Leaving the notion of the even briefer informational message of direct-to-consumer broadcasted advertisements, we will now briefly explore the arguments of philosophers for recommendations regarding what information should be given to patients in the patient–physician dialogue. Our categorization of information now moves from brief summary information to what courts and ethicists have referred to as *material information.*

Ethicists have approached the notion of information in the patient–physician relationship from the standpoint of manipulation of information, rather than from the standpoints of cognitive biases or of too much information. For example, Faden and Beauchamp focus on manipulation of information by physicians and behavioral aspects of how information is disclosed and discussed. Borrowing a term used in court decision making, Faden and Beauchamp emphasize that the condition of *substantial understanding* on the part of patients "demands apprehension of all the *material* or *important* descriptions." But the authors argue that what patients are willing to authorize and tolerate is "based not only on immediate beliefs and desires, but also on longstanding and long-range goals and values." The authors describe these beliefs and desires as "personal attributes" that, for the authors, "are complicated products of personal history and social environment." They characterize this substantial understanding of material descriptions as key for a patient simply on the basis of the value of understanding as well as the possibility that the descriptions could *cause* the patient to change his or her mind one way or another.[11]

The authors' focus on information is somewhat limited in scope. What is needed is a firm understanding of what information the patient brings to the patient–physician encounter, both from nonphysician sources and from other physicians. Faden and Beauchamp's focus on attention to the patients' values has been recognized by earlier ethicists, such as Robert Veatch.[12] Yet patients are continually evolving regarding what values they consider important, and this evolution influences the *apparent preferences* patients have for or against medical interventions based on short-, medium-, and long-term outcomes. Some patients, particularly older patients, patients with many medical conditions, and patients that have borne medical conditions and disabilities over time, may have *more stable preferences* than others. Patients who have neither borne medical conditions or diseases themselves nor seen similar morbidities borne by others may be at a particular disadvantage. These individuals may not only have *unstable preferences,* but also may be at a particular disadvantage when formulating their own preferences for or against medical interventions.

What ethicists need to explore is the role of information and its continued impact on patients' value considerations. Cognitive psychologists have systematically studied the impact of how framing information influences decision making in many hypothetical circumstances, yet that same information may determine the way patients understand and appreciate their own values.

Thus, a systematic exploration of how information is to be evaluated in decision making and value formulation is needed, not only by social scientists but also by philosophers.

The final concept that we see as being dominant in clinical research is the notion of *full disclosure*. Here the Belmont Report argues that "the research subject, being in essence a volunteer, may wish to know considerably more about risks gratuitously undertaken than do patients who deliver themselves into the hands of a clinician for needed care."[13] Thus a fuller disclosure is required of the individual (patient) being recruited into a clinical research study. Indeed, in recruiting an individual into a clinical research study that involves risk to humans, principal investigators are required by federal regulations to construct a written informed consent form, a copy of which goes with the patient. This written informed consent form itself must also be approved by an Institutional Review Board, which reviews in detail the scientific protocol, the ethical implications of the study for human participants, and the contents and language of the informed consent form.

Now we will go on to explore the domain of basic decision-making processes in clinical care and clinical research.

A. Basic Medical Decision-Making Processes

In a typical clinical encounter, the patient comes to the physician for an opinion on a clinical symptom or sign. The physician, on the basis of his or her medical training attempts to sort out and identify the cause or causes of the symptom or sign through testing and to develop a therapeutic or management plan.

Recent social scientists and philosophers have approached the patient–physician dialogue by focusing on what I consider a much too limited set of information. Let us first explore the full domain of medical decision making and its processes and then discuss how this book helps clarify that domain for social scientists and philosophers. The real problem is understanding both the *domains* of the medical decision-making process and the patient's and physician's *access to information* in these domains.

B. The Domains of Medical Decision Making

Graduate students—the future theoreticians, researchers, and writers—must have a solid foundation in what the medical decision-making process is about in each of three domains of medical decision making. Each domain is characterized by a patient and a provider. In most instances, the provider will be a primary care provider who may be a generalist (or a nurse practitioner or physician advocate or other CAM provider, e.g., a naturopath or chiropractor) and a patient. If the primary care provider elects to do a consultation to a subspecialist, a third party comes into the relationship. At this time, a general practitioner might give up the care of the patient to the subspecialist, or there might be a three-way patient–primary care provider–subspecialist relationship, with the potential for a three-way discussion. But rarely will all three participants be in the room at the same time. It is important to note that there will always be a potential problem with *access to the subspecialist.*

In clinical medicine and research, there are three basic domains:

Domain 1: the doctor's office or clinic office and general medical care (with extension to subspecialty care in select cases);

Domain 2: the medical ward in-hospital after one or more physicians (in primary care, subspecialty care, or emergency room) have elected to hospitalize the patient;

Domain 3: intensive medical care where the patient is first followed in an intensive care unit.

C. General Types of Decision Making across All Three Medical Domains

The medical decision-making process can easily be approached from four perspectives: those of clinical medicine, typical research medicine, atypical research medicine, and innovative therapies. Let us look at each separately.

1. *Clinical medicine* is the medicine that most patients see in their doctor's offices, in their HMO's clinics, in the medical wards, and in the intensive care units of medical centers.
2. *Typical research medicine* involves the recruitment of a competent patient from any of the above three domains into a research study approved by an Institutional Review Board to study an aspect of the patient's care, the patient's disease, or alternative ways of screening, diagnosing, and managing disease. Medical research obtains the most solid evidence from randomized controlled clinical trials, but because of cost, alternative research study designs are considered and evaluated.
3. *Atypical research medicine* involves research on newly developing issues, for example, genetic diseases. These are atypical because the full scope of potential problems has not been encountered as yet. The methodologies and protections for human subjects are still developing but are operative in certain medical centers.

4. *Innovative therapies* are therapies and management plans developed for patients who have not responded to more standard therapies, when there is no new research being done that could yield results expeditiously enough to help the patient. Therefore, purely on clinical grounds, a patient can embark on a therapy that, although approved for some diseases, has not as yet been approved for a particular clinic. Such innovative therapies do not have to go through an approval process by an Institutional Review Board but should be actively followed by the clinician.

We have examined the domains of medical decision making and the general types of decision making within each domain. It is now crucial to understand the following issues related to information and the patient–physician dialogue: Is information available? If not, when (if ever) is it expected to become available? What is the quality of that information? How is the information message presented to the potential patient? What is the format of the message? What is the intent of the developer of the information message? How can the message be improved? We are now ready to assess the area of information and the medical decision-making processes inherent in contemporary medicine and in the medical sciences.

Notes

1. D. Kahneman, P. Slovic, and A. Tversky, *Judgment Under Uncertainty: Heuristics and Biases* (New York: Cambridge University Press, 1982).

2. See K. Tates and L. Meeuwesen, "Doctor–Parent–Child Communication: A (Re)view of the Literature," *Social Science and Medicine* 52, no. 6 (March 2001): 839–51; B. M. Korsch, E. K. Gozzi, and V. Francis, "Gaps in Doctor–Patient Communication: I. Doctor–Patient Interaction and Patient Satisfaction," *Pediatrics* 42, no. 5 (November 1968): 855–71; and B. M. Korsch and V. F. Negrete, "Doctor–Patient Communication," *Scientific American* 227, no. 2 (August 1972): 66–74.

3. See C. Charles, A. Gafni, and T. Whelan, "Shared Decision-Making in the Medical Encounter: What Does It Mean? (Or It Takes at Least Two to Tango)," *Social Science and Medicine* 44, no. 5 (March 1997): 681–92, and "Decision-Making in the Physician–Patient Encounter: Revisiting the Shared Treatment Decision-Making Model," *Social Science and Medicine* 49, no. 5 (September 1999): 651–61; F. A. Stevenson et al., "Doctor–Patient Communication about Drugs: The Evidence for Shared Decision Making," *Social Science and Medicine* 50, no. 6 (March 2000): 829–40; and M. Gattellari, P. N. Butow, and M. H. N. Tattersall, "Sharing Decisions in Cancer Care," *Social Science and Medicine* 52, no. 12 (2001): 1865–78.

4. On cancer, see M. A. Richardson, "Biopharmacologic and Herbal Therapies for Cancer: Research Update from NCCAM," *Journal of Nutrition* 131, no. 11 suppl. (No-

vember 2001): 3037S–40S; on chronic pain, see S. Nayak et al., "The Use of Complementary and Alternative Therapies for Chronic Pain Following Spinal Cord Injury: A Pilot Survey," *Journal of Spinal Cord Medicine* 24, no. 1 (Spring 2001):54–62; on otolaryngology, see B. F. Asher, M. Seidman, and C. Snyderman, "Complementary and Alternative Medicine in Otolaryngology," *Laryngoscope* 111, no. 8 (August 2001): 1383–89.

5. See T. L. Beauchamp and J. F. Childress, *Principles of Biomedical Ethics*, 5th ed. (New York: Oxford University Press, 2001).

6. FDA Modernization Act of 1997 (Public Law 105–115, 105th Congress). Accessible online at www.fda.gov/guidance/105–115.htm-101k, accessed September 17, 2002.

7. Federal Register, August 12, 1997.

8. P. Maguire, "How Direct-to-Consumer Advertising Is Putting the Squeeze on Physicians," *ACP-ASIM Observer*, March 1999, at www.acponline.org/journals/news/mar99/squeeze.htm (accessed July 4, 2002).

9. Center for Medical Consumers, Inc., "FDA Relaxes Direct-to-Consumer Advertising Rules," *HealthFacts*, February 1998, at www.medicalconsumers.org (accessed December 1, 2001).

10. See R. R. Faden and T. L. Beauchamp, *A History and Theory of Informed Consent* (New York: Oxford University Press, 1986), 235–48, 276–81, 283–88.

11. Faden and Beauchamp, *History and Theory of Informed Consent*, 251–53, 278, 300–04.

12. R. M. Veatch, *The Patient as Partner: A Theory of Human-Experimental Ethics* (Bloomington: Indiana University Press, 1987), and *The Patient–Physician Relation: The Patient as Partner, Part 2* (Bloomington: Indiana University Press, 1991).

13. National Commission for the Protection of Human Subjects of Biomedical and Behavioral Research, *The Belmont Report: Ethical Guidelines for the Protection of Human Subjects of Research*, DHEW Publication, o.s., 78-0012 (Washington, D.C.: The Commission, 1978). Accessible online at http://ohsr.od.nih.gov/mpa/belmont.php3.

6

The New Medical Conversation
and the Scientific Information Message

NEW MEDICAL DRUGS and products and new medical technologies always raise compelling questions regarding information. But no less compelling are questions about the very nature of the information communicated about the drug, product, or technology, and the conversations that are held between patients and their doctors. We are concerned with the information that goes into and then gets digested in such conversations, which have gone through a history and evolution of their own in response to the courts, ethicists, communication researchers, and others. But as we shall see, these conversations are also constricted by two parameters. First, limited time is available. Second, there is the increasing number of alternatives available for patients and physicians to talk about.

Medical information was itself initially confined to the profession of medicine. But the Internet has changed this restriction forever. In addition, Congress has allowed product manufacturers to do direct-to-consumer advertising of prescription drugs. The marketing of drugs, products, and technologies often takes place with what appears to be too little regulation of the nature of the informational content of those messages. In addition, what scientific content there is in these messages is placed within an overall context that is designed to advertise the product, not to educate the consumer.

The range of formats and types of presentations of medical information to consumers and patients needs our focused attention, as does the ongoing debate surrounding what, when, and how information should be conveyed. The range of choices as to how to communicate medical information is very wide. On one extreme is the short advertised message promoting a medical drug or product. At the other extreme is the medical information conveyed in shared decision making, with or without the use of computerized decision support systems, where it is often argued that patients should have access to as much information as they want or need to make a decision.[1]

Short information messages are not confined to the promotional message designed by an advertising company about a new drug or medical product. Some are composed by panels of national and international experts, including physicians, scientists, statisticians, and epidemiologists. The experts are brought together often by government officials or other interested parties, in the attempt to develop a consensus view on a pressing health issue, such as the screening, diagnosis, or treatment of a particular medical condition causing morbidity or mortality within a population. This book maps and further explores medical information transmission to consumers and patients in terms of the responsibilities attached to the design of such consensus panel messages. We explore real and practical ways of helping all interested parties better understand the purposes and limitations of such communications, and we focus on the physician's dual roles as the learned intermediary between the advertisers and the patient and as the interpreter of the information message of the expert consensus panels and the patient. Our focus is on how information and information messages are designed and formulated in three areas: clinical care, screening for disease, and clinical research. Our particular attention is on risk and benefit in health and medical care, and how risk information is presented to patients and to the public at large. The information messages we deal with can be found, in a sense, everywhere: on billboards, on television, in journal and magazine advertisements, and in the newspaper.

There is a spectrum to the way these messages are delivered in each of the three sectors. At one extreme are those diseases that form the bulk of morbidity and mortality within a society, such as coronary artery disease in the Western Hemisphere. At the other are "orphan" or rare diseases. The European Union, for example, defines orphan or rare diseases as less than five cases per 10,000 people.[2] These are called "orphan" diseases because it is not commercially attractive for the pharmaceutical industry to produce treatments for such a small number of affected patients. An example of an orphan disease is Fabry's disease. Gillaine Arduin, executive director of the European Organization for Rare Disorders in Paris, acknowledges that "With a rare disease it is difficult to exert much political influence."[3] Since 1983, the United States has designated 1001 drugs for orphan diseases. Of these, 204 have actually reached the market.[4] Should this number be considered a success or a failure? The answer depends on whom you ask: patients with the rare disorders, family members of these patients, advocates of research, or others. But the rarity of these conditions and the rarity of therapeutic successes related to them need more attention.

Disease screening provides a unique perspective on the issue of risk communication, because disease screening overlaps with both clinical diagnosis and clinical research. Many of the diseases that are screened for are relatively common, and therefore get more attention. But the development of screening tests does not mean that adequate therapies are available, or even that adequate research has been conducted to determine whether a therapeutic intervention early in the disease is better than a "watchful waiting" strategy. Two examples of screenable diseases for which compelling research is lacking are breast cancer in women age 40–49 and prostate cancer in men younger than 65.

Screening for disease is done in patients who are asymptomatic but may be at risk for developing the disease. This raises key questions regarding falsely labeling patients with regard to a disease process. A patient can be wrongly labeled as being disease-free because of a false negative test result, or as having the disease because of a false positive result. Sensitivity and specificity of many tests are known now so that the risk of a false positive or a false negative may be estimated and communicated. But are these actually understood by a patient with the positive or negative test result? Furthermore, how are patients expected to react to the knowledge that compelling scientific evidence is lacking? Some may argue that this is a toss-up situation, and the patient should be allowed to proceed as he or she sees fit. But what if the patient wants to know what the experts would do if they were faced with such a condition? Thus, there are problems with interpretation of even common tests.[5]

Screening for asymptomatic disease can be a double-edged prospect for the patient. For example, a free blood test for elevated prostate-specific antigen (PSA) may be offered as an inducement to attract patients to a new clinic or hospital.[6] These free checks can be anticipated to yield a number of false positive results, which will need to be interpreted by a provider and followed up over time. So such patients will have to spend time in the clinic or hospital getting an interpretation of the result, or take the result back to their own physician for interpretation.

Pertinent issues for the patient may include the following:

- What is the risk of my test results being a false positive or false negative?
- Is there a "gold standard" test available as the next test I will undergo in my evaluation of this asymptomatic disease?
  - If so, what are its risks and benefits?
  - If not, what am I to do? What would the experts do in my circumstances?

• What are the currently available therapies, and what is the likelihood that they will have a meaningful impact on my health, my lifespan, and the quality of life in my remaining years?

Geyman, Deyo, and Ramsey note that different organizations may give quite divergent recommendations concerning screening.[7] In the absence of solid data, one group may simply report that absence, while another group may formulate a national or international consensus panel to provide recommendations. Another approach would be to list all approaches used by different organizations and then allow the provider or the patient to make a decision as to which organizations' recommendations to follow.

Each approach yields a different set of information to be evaluated. Let us examine three facts. The first is that clear and compelling screening evidence may not exist because exhaustive research studies have not been done. The second is that even expert consensus panels may not agree on a recommendation for or against screening. The third is that the absence of scientifically compelling data may be quite emotionally and intellectually disconcerting to the patient. These facts serve to highlight four questions:

• What is the purpose of the information message?
• What information should be contained in the information message?
• How is the information message designed and delivered to the patient, consumer, and citizen?
• Who should be in control of the parameters used in the design of the information message?

We now examine these four questions.

Question 1. What Is the Purpose of the Information Message?

The purpose of the information determines what is contained in the message. The information content will differ if the purpose is to promote a product or service as opposed to educating, instructing, or informing patients with a view to decision making.

Question 2. What Information Should Be
Contained in the Information Message?

This is perhaps the most important question, and generates another set of questions:

- How much time and space will be allowed for descriptions of the purpose of the product? What description should be used regarding who (which individuals and which groups) can use the product and with what risks and what benefits? Who cannot use the product because of particular medical conditions they have or particular drugs they are taking?
- Is the content geared solely for advertising purposes? If so, then what is the purpose of any scientific information the advertisement may contain?
- Is the content going to have "mixed purposes," for example, for education and advertising? If so, then who decides what is to count as education and what is to count as advertising? And what is the relative amount of time or space to be spent on educating vs. advertising?
- Finally, at whom is the content of the short information message aimed? Is it aimed at the scientist, the consumer with the medical condition, or the consumer with some but not all of the symptoms? And if individuals differ in the categories, types, and amount of information they need, then *at whom* is the message aimed?

Question 3. How Is the Information Message Designed and Delivered to the Patient, Consumer, and Citizen?

The question of how the information message is designed relates both to its purpose and to its information content. What "limits" should be set on the design of the message?

Let us take the case of *subliminal messages.* Can we all agree that subliminal messages should be eliminated from the information message of medical products, drugs and devices? Lewicki, Hill, and Czyzewska argue that these "mechanisms of non-conscious acquisition of information provide a major channel for the development of procedural knowledge." Procedural knowledge "is indispensable for such important aspects of cognitive functioning as encoding and interpretation of stimuli and the triggering of emotional reactions."[8] Perspectives on subliminal messaging appear in the literature of the educational as well as the forensic sciences.[9] In addition, in moving away from subliminal messages to spoken or printed information messages, it is often hard to distinguish what in fact communicates knowledge or expert opinion versus what is a marketing ploy.[10]

Question 4. Who Should Be in Control of the Parameters Used in the Design of the Information Message?

For years, the courts have placed the physician as a learned intermediary between the advertised message and the consumer in medical health care. Research in communication in disease screening, if done well, may in turn influence risk communication in both clinical diagnosis and clinical research. But the basic difficulty is understanding what information should go into the construction and design of the information message and why that information should be there. The aspects of *scientific description* and *scientific numeracy* that should appear in a short advertising message will depend on whether that message is designed and constructed for scientific purposes only, for advertising purposes only, or for both scientific and advertising purposes.

Selective Information Messages

The types of messages we will be considering are *selective information messages*, that is, messages containing a selective sample of all the known information about a medical condition or disease. These messages are communicated in various forms: sometimes as recommendations, sometimes as pronouncements, and sometimes as information to be incorporated into patients' informed decisions about their health and medical care.

Six issues may not be clear to the hearer or viewer of an information message:

1. For what purpose is the information designed?
2. Does scientific content actually appear in the information message, or is it simply alluded to in a reference to "experts"? This breaks down into two topics:
   - If scientific information (scientific data) is presented, what was the method used to select the data that are being presented and the data that are being left out? If there is an imbalance in the presentation of *benefit information* vs. *risk information*, why is this imbalance allowed to exist? And who should decide how any such imbalances should be rectified?
   - If there are no scientific data presented, but only a reference to experts, how were the experts selected? For example, are there any potential conflicts of interest involving the experts and the product being advertised?
3. Is it easy for the reader or hearer of the information message to separate out the "scientific data," the "experts' perspective," and the advertising aspects of the message?
4. How much input did relevant parties have into the design of the message?
5. Who is presenting the information message? Are there any conflicts of interest related to the presenter, and if so, what are they?
6. How simplistic is the scientific information being presented?

Frank Kee observes that

> a recent study has shown that it is not an increase in data given to patients per se that affects their preferences but rather the level of explanation of the data. . . . [W]hen patients express complex contingencies relevant to their treatment decisions, health care professionals should avoid simplifying those complexities into a list of pros and cons.[11]

The hearer of the information message may not be clear on the following points:

- Who actually developed and designed the message? Was it the company or product manufacturer?
- Was a slant placed on the message to obfuscate the scientific information presented, or to confuse the hearer regarding the exclusion of scientific information?

The hearer of an information message might ask, "Why is this information message being aimed at me, and why now?"

Anyone familiar with marketing research and advertising understands the factors that are scrutinized to achieve a message that will distinguish a product or a product manufacturer from its competitors. However, we must be extremely careful when the message being delivered is a short medical information message to the general public, a patient in a clinic or hospital, or an individual who is being approached as a potential research participant. Clearly, in these areas, the object of a short information message should not be merely to draw business away from a competitor. Many times the message is, in fact, geared toward drawing an individual away from a *competing perspective*.

Information messages are designed to influence decisions. These messages target citizens in the case of public health, patients in the case of clinical care, and individuals in the case of clinical research. We will be interested in three communication perspectives:

- How is risk vs. benefit information being communicated?
- How is the risk and benefit message designed to fit into an individual's decision-making framework?
- From whose perspective is the risk and benefit message designed and dispersed?

People often point out that all decisions involve risks. Some say that just getting up in the morning presents a risk. Yet even simple information messages may be misleading, in the sense that the designer of the message tries to put the individual into a frame of mind such that risk is, in some sense, acceptable.

It is important to recognize at the outset that risk may be viewed differently in different domains. Why? The stakeholders in the decision are viewed differently in each domain. The stakeholders in the domain of environmental health, industrial health, occupational health, and public health in general may be the entire population of a country or may be a smaller identifiable group within a population. The risk messages that come from industry, from government, and from experts in universities can be directed either at the public in general or—bypassing the public entirely—at providers that are involved in the health care, clinical care, and clinical research spheres.

Let us take the case of informed consent, an identifiable discrete area in medical tort law that is influenced both by judicial opinion and by decisions in state legislatures. The public does vote but does not sign informed consent forms. Individuals in clinical care and those individuals who elect to participate in clinical research do sign informed consent forms, accepting risks on the basis of disclosures made to them as citizens with specific legal rights related to medical and research interventions. Those legal rights include self-determination, autonomy, volunteer choice, and self-decision. However, even in the case of individual self-decision in medical care, there is constraint.

In most cases, patients cannot make decisions themselves regarding what they want in terms of medical care or medical research. Rather, they can refuse to accept medical interventions and research interventions that are offered to them. In addition, individuals can elect to change the physicians and institutions providing their care. But they cannot just have whatever they want in terms of clinical care and clinical research. Or so it seemed before direct-to-consumer advertising, which encourages patients to ask for certain prescription drugs. Many times their physicians will oblige. Phyllis Maguire notes that

> while the boom in direct-to-consumer advertising has been a bonanza for Wall Street and Madison Avenue, it has also sparked a surge in physician office visits. According to Scott-Levin, a drug marketing research firm in Newtown, Pa., while all office visits to doctors rose 2% during the first nine months of 1998, visits for conditions targeted by ad campaigns rose much more dramatically. Patient visits for smoking cessation rose 263%, for example, while visits to treat impotence jumped 113%, hair loss 30%, osteoporosis 22%, high cholesterol 19% and allergies 11%.[12]

Yet all patient or research participant self-decision must itself be understood in the context of other types of constraint. Most constraint comes in the form of cost, access to care, or access to clinical trials. But constraint can also involve attitudes, beliefs, and values held by individual providers, for example, physicians caring for the patients. These may best be considered privately held opinions, because (except in the case of emergency care) providers also have autonomy in terms of the individuals they care for and the standards of care they follow. But all of these privately held attitudes, beliefs, opinions, and values must be shared with all patients who explicitly request information, because information affects patient choice and decision making. Other patients may not want information and may be content to track how decisions are made on their behalf by others.[13] Still others may want to "test the waters" of decision making. In each case, the issues are when to provide information, how much information is wanted, and what formats of information will be most useful to patients.

While it may be difficult to search the medical literature to find what has or has not been done regarding a clinical issue, it is still a doable task. However, a provider's personal opinions or values can only be accessed in the main through conversations with patients. These value-oriented conversations are often the toughest part of decision making. They require both parties, patient and physician, to learn to share personal information and privately held beliefs. In addition, and perhaps most importantly, two questions arise on the part of the physician: when should privately held beliefs of the physician enter the information message, and how should such beliefs be identified as privately shared opinions, beliefs, and values?

The issues of how risk and benefit are to be communicated in clinical care, public health, and clinical research and of how risk and benefit are judged are crucially interrelated. The main reason for communication about risk and benefit is to allow persons to judge risk and benefit for themselves in the case of clinical care, and to judge risks borne by individuals and garnered for society in the case of public health.

How Information Is Presented vs. What Information Is Not Presented

Issues abound as to what choices should be made by patients, and the influences on those choices. First, how information is presented affects choice.[14] If a patient is presented information about the chance of survival at, say, the time of a surgical intervention and at one, two, three, four, and five years after the intervention, many patients will opt for the treatment with a better five-year outcome. If that same patient is presented the same information in terms of the chance of *dying* at the time of the intervention and at one, two, three, four, and five years after the intervention, many patients will opt for the treatment with the better short-term chance for survival. Yet survival and mortality data are simply the inverse of each other: if 90 percent of patients survive, 10 percent of patients die; if 10 percent of patients survive, 90 percent die. Yet, the presentation of data as *survival data* or *mortality data* changes patient decision making.

Second, decision making in medicine conducted in doctor's offices, clinics, and hospital wards is *constrained by available time*. This limits the amount of information that is presented, the detail in which that information is presented, and the opportunity for questions. Thus, not all relevant information about risk, benefit, and presentation effects of information can all be given equal time and weight. What information does not make the cut? Indeed, if the patient wants systematic information about quality of life, such research may not even have been done as yet.

However, the problem in all three areas—clinical care, public health, and clinical research—occurs when the risk-benefit information is not delivered neutrally, but is delivered along with distracting information, such as marketing or advertising information. Although risk-benefit communication is realistic as a goal, the three frameworks we are interested in have their own historical origins and separate evolution. And each perspective has focused, primarily, not on risk-benefit communication, but on risk communication.

In clinical care, risk-benefit communication occurs within the patient–physician relationship. The earliest court cases were based on the understanding that physicians sought their patients' consent prior to a medical intervention. This evolved into the judicial notions of *consent* and *informed consent*. While state legislatures have in some ways attempted to influence informed consent, it is the federal and state courts that have refined the concepts so far.[15]

Roger Kasperson and Pieter Jan Stallen trace risk analysis back at least to the Babylonians in 3,200 B.C.

> [C]ultures have traditionally used a host of mechanisms for anticipating, responding to, and communicating about hazards—as in food avoidance, taboos, stigma of persons and places, myths, migration, etc. . . . Seals at sites of the ninth century B.C. Harappan civilization of South Asia record the owner and/or contents of the containers. The Pure Food and Drug Act, the first labeling law with national scope in the United States, was passed in 1906. . . . In any specific setting, the different interested groups are likely to have different motivations, which makes risk communication, whether it is at the national or local level, an essentially political event.[16]

In public health, Kasperson and Stallen note that "the bulk of risk communication research conducted during the 1980s has been driven by the pragmatic needs of government and industry."[17] For the authors, the prevailing conception of risk communication is

- predicated upon the assumption of a neutral, altruistic communicator;
- heavily oriented towards the product (the understanding of the message) as opposed to the process (developing an enduring capability for handling uncertainty in those potentially at risk);

- focused on the goals of the communicator;
- structured according to a communications engineering model involving senders, media, messages, and receivers.

Kasperson and Stallen argue that both the communications engineering approach and the advertising approach are intrinsically inadequate for the design, analysis, or evaluation of risk communication programs. They note that "the terminology of discourse about risk communication identifying 'sources,' 'channels,' 'messages,' 'audiences,' and 'targets' is extraordinarily revealing about how the risk communication process is conceived."[18] The authors find that risk communication is a highly contentious value-setting activity potentially associated with both positive and negative consequences:

- positive consequences of warning about hazards (such as self-protection), but with potential unintended adverse consequences (such as unintended worry, interpersonal conflict, and risk-enlargement);
- potential increased self-protection of the individual, but at the cost of erosion of personal autonomy;
- the goal of public interest, but with potential self-interest and bias of individuals (such as individuals reflecting biases of past education, professionalism, relations with superiors in an organization, and goals and missions of an institution);
- real compassion and caring, but with the potential of "a veneer of words that fail to correspond to actual values and concerns";
- proceeding with communication about risk, but in the absence of rigorous evaluation of that communication.[19]

Interestingly, Kasperson and Stallen believe that the underlying approach used by communications engineering has been augmented by experience from advertising and marketing that is intellectually compatible in that it too proceeds from a base that assumes the following four parts:

- the intentionality of the communication process;
- the flow of information from the communicator to the target as the dominant communication task;
- the creation of the described behavior as the outcomes by which the success of the communication effort should be judged;
- the view that the key to effective communication rests upon selection of the most appropriate technique as geared to various target recipients.[20]

The authors are drawn into risk communication as an important part of more general risk management. But they argue that people have to get through the barriers that exist in risk communication to get to the final discussions about whether these risks and benefits are at all reasonable, either individually or with respect to the population.

It is important to recognize that risk and benefit can be intentionally shaped to point people in a particular direction, rather than to allow neutral interpretation and judgment. And this intentional shaping and influencing of people becomes a crucial focus of the message and the values on which it is based. We will examine the potential importance of *value-based messages* that attempt to communicate risk-based information on specific sets of values and beliefs.

The basic issue here relates to the fact that someone cannot judge risk and benefit without a set of fundamental tools on which any judgments are grounded. First, he or she must have *relevant information.* Yet, information may be incomplete and far off in terms of its scientific discoverability. Second, he or she needs information that is *translated.* Information of interest in risk and benefit judgment comes from technical sources, like scientific information and various forms of statistical and probabilistic information. Third, this translated information must be *communicated* to the individual. Such communications may take various forms, including brief advertising messages found on television, more complex opinion articles that appear in the medical scientific literature, and less complex information presented in consumer journals and newsletters. Fourth, the communicated information must be *valued* and *evaluated.* That is, scientific information—in particular, medical scientific information—is in most senses value-neutral if it is presented in its entirety with attention to the experts' level of confidence in the quality of data and its interpretation. Yet this scientific information must subsequently be placed in different value systems before decision making can occur at the level of the person or society. Value here includes the scientific worth of the information in light of competing theories and of how that scientific information is interpreted. The subsequent uses of scientific information then become value-laden, as information is summarized, interpreted, and translated into language that can be understood by the nonscientist. This value-laden component of scientific communication is the topic of the next chapters.

Notes

1. See D. J. Mazur, *Shared Decision Making in the Patient–Physician Relationship: Challenges Facing Patients, Physicians, and Medical Institutions* (Tampa, Fla.: American College of Physician Executives, 2001); and A. Edwards and G. Elwyn, *Evidence-Based Patient Choice: Inevitable or Impossible?* (Oxford: Oxford University Press, 2001).

2. W. Weber, "European Parliament Raises Awareness for Patients with Rare Diseases," *Lancet* 356 (October 28, 2000): 9240.

3. Gillaine Arduin, quoted in Weber, "European Parliament Raises Awareness," 1504.

4. Weber, "European Parliament Raises Awareness," 1504.

5. See D. Ransohoff and A. Feinstein, "Problems of Spectrum and Bias in Evaluating the Efficacy of Diagnostic Tests," *New England Journal of Medicine* 299, no. 17 (October 26, 1978): 926–30; and J. G. Elmore et al., "Ten-Year Risk of False Positive Screening Mammograms and Clinical Breast Examinations," *New England Journal of Medicine* 338, no. 16 (April 16, 1998): 1089–96.

6. See D. J. Mazur, *Shared Decision Making in the Patient–Physician Relationship: Challenges Facing Patients, Physicians, and Medical Institutions* (Tampa, Fla.: American College of Physician Executives, 2001), 131–34, 294–95.

7. J. P. Geyman, R. A. Deyo, and S. D. Ramsey, eds., *Evidence-Based Clinical Practice: Concepts and Approaches* (Woburn, Mass.: Butterworth-Heinemann, 2000), 92–93.

8. P. Lewicki, T. Hill, and M. Czyzewska. "Nonconscious Acquisition of Information," *American Psychologist* 47, no. 6 (June 1992): 796–801, quoted from p. 796. See also J. de Houwer, J. H. Hendrickx, and F. Baeyens, "Evaluative Learning with 'Subliminally' Presented Stimuli." *Consciousness and Cognition* 6, no. 1 (March 1997): 87–107.

9. See S. A. Henry and R. G. Swartz, "Enhancing Healthcare Education with Accelerated Learning Techniques," *Journal of Nursing Staff Development* 11, no. 1 (January/February 1995): 21–24; and R. E. Litman and N. L. Farberow, "Pop-Rock Music as Precipitating Cause in Youth Suicide," *Journal of Forensic Sciences* 39, no. 2 (March 1994): 494–99.

10. L. J. Peterson, "Innovation or Marketing Ploy: How Do We Distinguish between the Two?" *Oral Surgery, Oral Medicine, Oral Pathology, Oral Radiology, and Endodontics* 92, no. 1 (July 2001): 1.

11. F. Kee, "Patients' Prerogatives and Perceptions of Benefit," *British Medical Journal* 312 (April 13, 1996): 958–60, quote from p. 959.

12. P. Maguire, "How Direct-to-Consumer Advertising Is Putting the Squeeze on Physicians," *ACP-ASIM Observer*, March 1999, at www.acponline.org/journals/news/mar99/squeeze.htm (accessed July 4, 2002).

13. Mazur, *Shared Decision Making*.

14. See B. J. McNeil et al., "On the Elicitation of Preferences for Alternative Therapies," *New England Journal of Medicine* 306, no. 21 (May 27, 1982): 1259–62.

15. See S. Jasanoff, *Science at the Bar: Law, Science, and Technology in America* (Cambridge: Harvard University Press, 1995).

16. R. E. Kasperson and P. J. M. Stallen, "Risk Communication: The Evolution of Attempts," in *Communicating Risks to the Public: International Perspectives,* ed. Kasperson and Stallen (Boston: Kluwer, 1991), 1–11, quoted from p. 1.

17. Kasperson and Stallen, "Risk Communication," 2.

18. Kasperson and Stallen, "Risk Communication," 4.

19. Kasperson and Stallen, "Risk Communication," 8–9.

20. Kasperson and Stallen, "Risk Communication," 5.

7

The Circumscription of Information by the Courts

H ISTORICALLY, THE FIRST INFORMATION MESSAGES that were evaluated by courts were simple acts by physicians of securing their patients' consent to treatment. Ironically, the courts were asked to consider questions related to medical interventions that were not even necessarily associated with particular information messages. The basic question considered by these early courts was whether a physician must secure a patient's consent prior to a medical intervention. In these first cases, the courts were more concerned that the physician secured the consent of the patient prior to the intervention and less interested in what information was used in securing that consent.

The first case of consent was a combination case in which a physician took a fracture of a femoral bone that was healing with callus formation, rebroke it, and then set it in a mechanical device with teeth. Thus, this case had elements of experimentation as well as clinical care. The court was interested in whether it was part of the standard of care of physicians as professionals to secure a patient's consent prior to a medical intervention. Here, the court held that the right of trespass on a patient's body would not be permitted unless the physician had obtained the patient's consent prior to the intervention.

Even in the early 1900s, U.S. courts heard cases in which patients, in preintervention discussions, laid down their beliefs in relation to what they permitted physicians to do in a medical intervention. For example, one patient specified that a physician could surgically explore a foot that was causing persistent discomfort but not remove any bones without explicit consent. The physician then agreed to that scope of intervention, but during the actual surgical procedure—and absent any emergency—the physician exceeded that scope and removed a bone. The physician argued that this was done on the basis of his medical judgment, but the court focused the argument on the fact that the removal of the bone in the absence of any emergency was done against the patient's expressed wishes. Thus, the physician's removal of the bone from the patient's foot exceeded what the patient had agreed to in absence of any emergency, and the court found for the patient.[1] Here the court argued that the physician had to discuss the scope of the medical intervention with the patient prior to the intervention, and any discrepancies between what the physician advised and what the patient considered a personally acceptable intervention had to be worked out before the medical intervention.

While the issue of *scope of an intervention* plagued physicians, judges, and juries in the early 1900s in the United States, the issue was one of obtaining the consent of a competent adult patient. Indeed, from the earliest consent case in Great Britain in 1767 until the term *informed consent* was first introduced into the judicial lexicon in a written court opinion in California in 1957,[2] there was only one standard that courts used to determine whether a physician had secured a patient's consent to a medical intervention, the *professional standard*. Under the professional standard, a physician had to disclose what a physician's community of peers deemed appropriate.

The notion that there should be a non-professionally defined standard in informed consent did not surface in judicial decision making until the early 1970s. In 1972 Judge Spottswood Robinson, in the landmark federal decision *Canterbury v. Spence*, created the *reasonable person standard of disclosure*.[3] Here a physician became obligated to disclose what a reasonable person in the patient's position would want to know. The two dominant judicial standards of informed consent—the professional standard and the reasonable person standard—were themselves based on nonscientific grounds. Courts based their decisions primarily on two sets of issues: precedent cases and the testimony of physician experts called into court by both parties (patient and physician). In the written court opinions, judges may cite the statistics provided by those respective physician experts that relate to the adverse outcome that occurred in the patient's case. In *Canterbury v. Spence*, Judge Robinson held that had the patient received the information about the adverse outcome, the patient would not have agreed to the medical intervention.

Judge Robinson did have a framework of categories that captured, for him, the essence of what the courts of his day were focusing on regarding medical information that needed to be disclosed to patients. The dominant category of information to be disclosed by a physician is *risk information*.

The second category is *benefit* information, the third a description of the *nature (and scope) of the procedure*, and the final category relates to *time* and whether the procedure can be delayed. Judge Robinson also recognized one of the biases in court deliberations about adverse outcomes that occur in medical interventions: *hindsight bias*. In hindsight, a patient who sustains an adverse outcome may have difficulties looking back at the informed consent session that preceded the decision to intervene from an objective perspective.

Many of the issues considered by U.S. courts in the early 1900s remain with us today. Of particular relevance are issues regarding the *scope of a medical intervention*. Often the noninvasive and invasive diagnostic testing that is undertaken prior to a surgical intervention may not be definitive enough to pinpoint the patient's problem. That is, the initial noninvasive and invasive testing may not reveal what the surgeon should expect to find upon biopsying an area with a patient under general anesthesia, or upon opening the abdomen. Such intervention requires an extensive discussion of what a patient wants the surgeon to do across a range of possible outcomes. This is perhaps best illustrated by the example of a woman who needs to go under general anesthesia for a biopsy of a breast mass. Does the patient just want the surgeon to do the breast biopsy and then recover from the anesthesia, hear the biopsy results, and then discuss alternatives? Does the patient want to discuss all options related to each possible biopsy result prior to the initial biopsy so that she will come out of the anesthesia with the intervention completed? Or does the patient want to designate a *significant other* and then have the surgeon come out of the operating room and discuss the options with the *designated decision maker* in the patient's absence?

Early courts recognized that physicians need to intervene medically with patients in settings not amenable to discourse and the discussion of alternatives. Thus, emergency cases were separated from the rules of informed consent. Questions today arise with the notion of urgent care and how urgency relates to emergency vs. elective interventions.

From its earliest beginning in the late 1700s, the notion of *consent* was inextricably linked to a competent adult patient's right to refuse medical interventions. The failure of a physician to obtain the patient's consent constitutes the tort of battery. This basic right to refuse medical interventions is reserved for the competent patient but may be exerted through documents stating the patient's wishes in certain circumstances. Yet much discussion of competency in decision making takes place in the absence of any clear discussion of "how much impairment we as a society are willing to tolerate before we consider someone incompetent."[4]

In the absence of a societal consensus, four elements have been argued by Appelbaum and others to be "the most commonly used standards for competency": the abilities to understand, to appreciate, to reason, and to evidence a choice.[5]

Our main focus is on the first ability, to understand. What does it mean to *understand*? How is understanding evidenced? And what is to be understood?

Understanding is most typically evidenced by *recall*, that is, asking patients to describe the information that has been presented to them in clinical care and clinical research. But Paul Appelbaum has argued that

> it is difficult to avoid the conclusion that psychiatric disorders, which by definition affect mentation, raise special concerns. Indeed, a recent study of decisional capacities in the inpatient treatment context demonstrated substantial impairments in approximately 52% of schizophrenic subjects and 24% of depressed subjects, but only 12% of seriously medically ill subjects. These data call into question . . . the ability of these subjects to protect their own interests in making decisions about entering research projects.[6]

In the case of psychiatric research in patients with schizophrenia, Thomas Posever and Theodore Chelmow propose a tripartite consent process that "calls for oversight of consent by three professionals, each with a distinct domain of responsibility and expertise and each with the authority to deny or accept a prospective subject's choice to participate in the proposed research." The authors argue that this tripartite review "incorporates independent assessment of capacity within the informed consent process through a rigorous system of 'checks and balances.'"[7]

> The process begins with an evaluation by the prospective subject's treatment team to assess whether an individual should be approached to participate in a particular study at all. If the treatment team believes that participation will not unduly compromise their patient's safety, the process of seeking his or her consent may go forward. It then continues with the employment of an appropriate consent form and an initial conversation between researcher and subject that results in provisional consent. Subsequent, independent assessments of a prospective subject's decision making capacity and exercise of free and considered choice assure that only those individuals who are able to give appropriately informed and voluntary consent will actually be enrolled in research.[8]

The authors describe this tripartite informed consent process as intended for "high-risk studies on schizophrenia."[9] But the question remains: What is to be understood about interventions and non-interventions in clinical care and clinical research? The judicial standard of informed consent was built on consideration of cases that mainly involved a patient's decision to undergo or refuse to undergo a medical intervention recommended by his or her physician. The judicial perspective thus focused on disclosures by the physician regarding the recommended intervention. The types of information courts wanted disclosed about this physician-recommended medical intervention included the nature of the procedure, its risks, and any reasonable alternatives open to the patient. As we have discussed before, the information disclosed is conditioned by the standard of informed consent that exists in the particular state—the professional standard or the reasonable person standard.

In the case of informed consent in clinical research, the Belmont Report created a specific standard for the clinical research setting involving human subjects: the *reasonable volunteer* standard. The Belmont Report focused on providing guidance related to the extent and nature of information to be provided to the "reasonable volunteer" in clinical research. It clearly distinguished the participant in clinical research from the patient being evaluated in a nonresearch clinical setting, stating that "subjects should understand clearly the range of risk and the voluntary nature of participation [in clinical research]." It is also important to recognize the study participants' right to discontinue their participation in a clinical research study without any untoward consequences and without any loss of the participants' privileges.

After discussing the notion of information, the Belmont Report makes key points about *comprehension*, that is, whether and to what extent the information is grasped by the individual being recruited. The first issue addressed in the report is the potential impact of the *manner* and *context* in which information is conveyed. Because a "subject's ability to understand is a function of intelligence, rationality, maturity, and language, it is necessary to adapt the presentation of the information to the subject's capacities." While the Belmont Report recognizes that the information provided to individuals being recruited into research studies "must be complete and adequately comprehended," it only offers the suggestion that "on occasion, it may be suitable to give some oral or written tests of comprehension." In cases where comprehension is *severely limited* (e.g., by conditions of immaturity or mental disability), the report states that "respect for persons also requires seeking the permission of other parties in order to protect the subjects from harm" and recommends that "the third parties chosen should be those who are most likely to understand the incompetent subject's situation and to act in that person's best interest." In addition, "the person authorized to act on behalf of the subject should be given an opportunity to observe the research as it proceeds in order to be able to withdraw the subject from the research, if such action appears in the subject's best interest."[10]

The Belmont Report holds that "an agreement to participate in research constitutes a valid consent only if voluntarily given. . . . [T]his element of informed consent requires conditions free of coercion and undue influence."

Coercion, per the Belmont Report, occurs "when an overt threat of harm is intentionally presented by one person to another in order to obtain compliance."

Undue influence occurs "through an offer of an excessive, unwarranted, inappropriate, or improper reward or other overture in order to obtain compliance."

Undue influence in the case of vulnerable individuals is also recognized by the Belmont Report. Here it is argued that "inducements that would ordinarily be acceptable may become undue influences if the subject is especially vulnerable." Examples of such influence include "manipulating a person's choice through the controlling influence of a close relative and threatening to withdraw health care services to which an individual would otherwise be entitled."[11]

In clinical research, there is a much broader specification of different categories of information that must be communicated to the individual being recruited. First of all, no one is ever compelled to participate in a research study or clinical trial on human subjects. Second, there should be a complete disclosure of the risks, common side effects, and possible severe outcomes, and of the following facts:

- The subject can terminate his or her participation in research at any time without penalty.
- The subject will not necessarily be compensated for injuries, but must sue to recover damages.
- A key individual in the research study may be contacted at any time.
- There may be unexpected outcomes.
- There may not be any benefit to the individual participating in the research trial.
- Research is not individualized care.

The more complete nature of the disclosures required in clinical research on human subjects is the dominant theme that distinguishes informed consent in clinical research from informed consent in clinical care.

Special Considerations in Informedness

Placebo-based research in clinical trials on human subjects deserves special consideration, because the subject must understand not only that the research does not involve his or her individualized care, but also that there is a specific random chance that he or she will get a placebo, that is, a biologically inert substance without intrinsic activity in relation to a disease process. Thus, a patient must understand (1) the fact of randomization, (2) the numerical chance of getting a placebo, and (3) the fact that a placebo has no intrinsic biologic activity of its own. The use of placebos in research on diseases for which there is existing therapy in standard clinical care is an issue of current and ongoing debate.[12] Much of this debate is, as Appelbaum has suggested, due to society's unwillingness to make a determination of the level of impairment necessary for a determination of incompetence. For example, researchers participating in the tripartite consent process may differ considerably in the amount of impairment they are willing to tolerate in a patient they are evaluating for competency to participate in a psychiatric research trial involving schizophrenia. Appelbaum notes that "instruments are being field tested, and we think they hold some hope for standardizing competence assessments. . . . The instruments themselves, however, only measure the degree of impairment in capacities."[13]

Let us go back to the competent adult patients who have been the main concern of judicial decision making. What happens when these competent adult patients are unconscious and unable to express a preference of choice? When an otherwise competent patient is rendered unconscious, the issue arises whether that patient has previously specified values to be held. The Ontario Court of Appeals argued that the right to make one's own medical decisions is an important freedom and for "this freedom to be meaningful, people must have the right to make choices that accord with their own values." In this court case, a patient was taken unconscious to a hospital and, in her personal effects, the nursing staff found a card that was signed by the patient but not dated. The card stated that the patient was a Jehovah's Witness and that under no circumstances was the patient to be given blood or blood products.[14] The court argued that "the no-blood card" spoke for the patient in absence of any evidence that she had changed her mind after signing the card.[15]

In this case, there is a set of questions involving scientific evidence that should have gone through the physician's mind when he gave the patient blood. First, was the signature in fact that of the patient's? Second, was there confirmatory evidence that the physicians could rely on that the patient indeed held the belief that was contained on the card? For example, were there members of the patient's family present who could vouch that the patient in question indeed had held the belief in question? And were these family members without a conflict of interest regarding the patient's non-treatment? Third, had someone else signed the card and dropped the card in her personal effects on the way to the hospital? And, fourth, was there time for the physician to conduct a meaningful inquiry of the patient's family or significant others to get answers to the above questions?

Due Care of Physicians and Risk Disclosure

Judge Robinson starts with the fiduciary duties of a physician based on trust as contrasted with the law of contracts (arm's-length transactions), noting that the physician is under an obligation to communicate specific information to patients for the following reasons:

- to alert the patient to symptoms that could pertain to a bodily abnormality;
- to tell the patient when his or her ailment is not responding to treatment;
- to instruct the patient about any limitations to be currently observed for his or her own welfare;
- to instruct the patient as to any precautionary therapy he or she should seek;
- to advise the patient of the need for, or desirability of, any alternative treatment promising greater benefits than that being pursued;
- to warn the patient of any risk to his or her well-being that the contemplated therapy might involve.

For Judge Robinson, "the patient's reliance upon the physician is a trust of the kind which traditionally has exacted obligations beyond those associated with arm's-length transactions." The fiducial quality of the patient–physician relationship thus requires the "duty of reasonable disclosure of the choices with respect to proposed therapy and the dangers inherently and potentially involved."[16]

The adoption of the *reasonable person* standard by the Canadian Supreme Court was a change in Canada, which previously had followed the professional standard. It is intriguing to look at the development of the reasonable person standard in Canada to see how the emphases of the doctrine of informed consent shifted. In the United States, informed consent is heard in terms of negligence, unless there is intentional physician misrepresentation of a procedure, fraud, or like claims, which are heard in terms of battery. In the landmark Canadian Supreme Court case, *Reibl v. Hughes,* the patient sustained a massive stroke, causing right-sided body paralysis and impotence. According to the *Reibl* opinion, the patient, prior to the procedure, queried the physician about the possibility of stroke. The surgeon did not inform the patient of the chance of being paralyzed during or shortly after the procedure but rather stressed that "the chances of paralysis were greater if the patient did not undergo the surgery."

Court deliberations are often based on the testimony of experts brought into the courtroom by the opposing sides in the case. Court decision making is based on an adversarial approach. Stephen Fienberg and the Panel on Statistical Assessments argue that court opinions attempt to present both the give and take of the adversarial process and due consideration of the qualifications of the experts "as a partial means of sorting out the evidence."[17] Yet the experts who appear in court are preselected by each side's attorneys. Attorneys can go through as many potential expert witnesses as they want, and some clients can afford to question many experts before they select those with the *right opinions* in a particular case. Indeed, experts in a field may differ widely in their opinions on a particular case; some may not even have an opinion. But one never hears the full range of expert opinion in a case.

The trial judge examined the issue of judicial precedents in Canadian law and based his decision on the *professional standard,* that is, "the duty of the surgeon as defined by accepted general practice in the neurosurgical community." On the basis of the expert testimony of two physician witnesses, the judge determined that the surgeon's duty to warn the patient does not extend to all the dangers incident to or possible in any surgical procedure, such as the dangers of anesthesia or the risk of infection, matters which men of ordinary knowledge are presumed to appreciate. The judge held that:

> I find, on the basis of the expert evidence, that in the circumstances of this case, the duty of the surgeon as defined by accepted general practice in the neurosurgical community was to explain to the patient:
> - the problem presented by stenosis [narrowing] in the specific artery in question . . .
> - the specific risks inherent in arterial surgery of this kind . . .
> - [and] the risks of continuing without surgery.

The judge added that "although elective surgery was indicated for the condition . . . there was no emergency in the sense that immediate surgical treatment was imperative."[18]

Again, the *Reibl* trial court heard the case under the professional standard. In the trial judge's opinion, "the plaintiff [patient] would flatly have refused" the intervention had the defendant neurosurgeon explained the risks as follows:

> I propose to remove a partial plug in an artery a few inches from your brain. There is a risk that as a result a fragment of tissue may slip into your brain and if it does, you have a 4 per cent chance of dying and a further 10 per cent chance of having a stroke.

The trial judge further argued that

> the defendant did not take sufficient care to convey to the plaintiff and assure that the plaintiff understood the gravity, nature, and extent of risk specifically attendant on the endarterectomy, in particular the risk that as a result of the operation he could die or suffer a stroke of varying degrees of severity. . . . [T]he single relevant area of concern was the relative likelihood of a healthy existence in the coming years with, as opposed to without, the surgery. I find that he did not address with the attention required of him the specific risks of an adverse result of the operation itself. The plaintiff was left with the impression that the operation carried no risks of consequence, other than those incidental to a surgical procedure.[19]

J. A. Brooke, speaking for the majority of the Ontario Court of Appeal, argued that "having regard for the emphasis which the learned trial judge places upon the statistical details, he has misunderstood the real significance of the evidence of [the two expert neurosurgeons]." Brooke noted that only one of the physician experts "made reference to statistics in addressing the manner in which he would advise his patient when seeking a consent to perform this operation. . . ."[20]

Let us review the numbers presented at trial. The defendant neurosurgeon put the chance of "death because of surgery" at 4 percent and the chance of "stroke because of the above surgery" at 10 percent, while one of the two physician experts put the chance of "death and stroke because of the above surgery" at between 2 and 4 percent.

Brooke states that the numbers provided by the defendant physician and the expert witness

> were really very different . . . [and] the reason for the difference went unexplained. No one asked the doctors. . . . If the difference is based solely or partly on the personal experience of the surgeons, and there is in the evidence some reason suggested that this may be so, then perhaps the explanation lies in the nature of the cases that each has dealt with and that the chance of survivorship of those undertaken by one was less than the other.
>
> If this is so, there may have been good reason not to mention statistics to the patient, but rather to simply contrast his position if he undertakes the surgery with that of not undertaking it and urge him to proceed because of his youth and

strength giving some assurance of survivorship. . . . [I]n my view, statistics can be very misleading. The manner in which the nature and degree of risk is explained to a particular patient is better left to the judgment of the doctor dealing with the man before him. Its adequacy can be simply tested.[21]

Speaking for the Supreme Court of Canada, C. J. Laskin found that in reaching this conclusion "the Ontario Court of Appeal went too far."

In my opinion, the record of evidence amply justifies the trial judge's finding that the plaintiff was told no more or understood no more than that he would be better off to have the operation than not to have it. This was not an adequate, not a sufficient disclosure of the risk attendant upon the operation itself, a risk well appreciated by the defendant in view of his own experience that of the sixty to seventy such operations that he had previously performed, eight to ten resulted in the death of the patients. Although the mortality rate was falling by 1970, the morbidity (the sickness or disease) rate . . . was still about ten per cent. The trial judge was also justified in finding that the plaintiff, who was concerned about his continuing headaches and who was found to be suffering from hypertension, had the impression that the surgery would alleviate his headaches and hypertension so that he could carry on with his job. [The defendant surgeon] made it plain in his evidence that the surgery would not cure the headaches but did not, as the trial judge found, make this plain to the plaintiff.[22]

In addition, Laskin argued that the following facts were relevant to the issue of whether a reasonable person in the patient's position would have declined surgery at the particular time:

- The patient was within about one and one-half years of earning pension benefits if he continued at his job.
- There was no neurological deficit then apparent.
- There was no immediate emergency making the surgery imperative.
- There was a grave risk of a stroke or worse during or as a result of the operation, while the risk of a stroke without it was in the future, with no precise time that could be fixed.

While Laskin's written opinion made clear the information that was required for disclosure to a reasonable person, other authors have not attended to Laskin's explanation of what a reasonable person would want in terms of clear communication of information from his or her physician.

Let us look at an example from Ellen Picard and Gerald Robertson's *Legal Liability of Doctors and Hospitals in Canada*. Picard and Robertson argue that

Reibl firmly establishes that doctors have a legal duty to advise their patients of the material risk (and other material information) associated with proposed treatment prior to obtaining consent. However, in *Reibl*, the Supreme Court of Canada also held that failure to provide this information to the patient may have rendered the doctor liable in negligence, but it does not vitiate the consent so as to make the doctor liable in battery.[23]

Here the authors conclude that "'uninformed' consent [amounting to *negligence*] does not mean 'invalid' consent [amounting to *battery*]." In addition, Picard and Robertson maintain that

> the amount of information to satisfy this test [that the patient must know what it is he or she is consenting to] is fairly minimal. . . . So long as the patient is told the basic nature of the proposed treatment, the consent will be sufficiently 'informed' to be valid (although . . . the doctor may still be liable in negligence if possible risks and other material information have not been disclosed to the patient).[24]

Here, the compelling issue for Picard and Robertson is that Canadian courts have to this time held that the underlying question is what constitutes knowledge of the *basic nature of the treatment*. If the case involved the patient not being informed of the basic nature of the treatment—for instance, through intentional misrepresentation by the physician—then the physician could be found guilty of battery. For Picard and Robertson, had the physician not provided what they describe as "collateral information" (e.g., about the risks of the proposed treatment), the result would not be in a finding of battery, but rather in negligence.

The Supreme Court of Canada in Reibl quoted from a precedent Supreme Court of Canada case, *Hopp v. Lepp* (1980).

> Even if a certain risk is a mere possibility that ordinarily need not be disclosed, yet if its occurrence carries serious consequences, as for example, paralysis, or even death, it should be regarded as a material risk requiring disclosure. The risk attending the plaintiff's surgery or its immediate aftermath was the risk of a stroke, of paralysis and of death.[25]

The adverse outcome of stroke attending surgery or its immediate aftermath in the *Reibl* case was estimated at between 2 and 10 percent, not just "a mere possibility." This is why we have examined *Reibl* before *Canterbury v. Spence*. It is not clear how the determination in the former case would have been made had the chance of stroke been 1 percent, 0.1 percent, or 0.01 percent.

The evolution of the concept of informed consent in Canada continued to involve the distinction between battery and negligence. The Supreme Court of Canada considered the failure to disclose attendant risks of a surgery as negligence on the part of a physician, not as battery.

The informed consent doctrine on which Canadian law is based is the reasonable person standard established by Judge Robinson in *Canterbury v. Spence.* The *Canterbury* decision was made at a time in the United States when the failure to disclose specific risks attendant on medical interventions was considered negligence. Judge Robinson's opinion instead rested on one key information and communication element: risk disclosure. In addition, Judge Robinson based his opinion not only on the severity of an adverse outcome but also on the probability of such an outcome occurring. Although Judge Robinson concludes that severe adverse outcomes like death and stroke need to be disclosed even as possibilities, he reviewed the chance of occurrence based on previous court decisions in developing his written opinion in *Canterbury.*

Regarding the scope of the information to be disclosed, Judge Robinson argued that "all risks potentially affecting [the patient's] decision must be unmasked." However, he also recognized that the physician cannot "second-guess the patient, whose ideas on materiality could hardly be known to the physician."

[The] physician's liability for nondisclosure is to be determined on the basis of foresight, not hindsight; no less than any other aspect of negligence, the issue on nondisclosure must be approached from the viewpoint of the reasonableness of the physician's divulgence in terms of what he or she should know to be the patient's informational needs. . . . [A] risk is thus material when a reasonable person, in what the physician knows or should know to be the patient's position, would be likely to attach significance to the risk or cluster of risks in deciding whether or not to forego the proposed therapy.[26]

Judge Robinson paid special attention to severe adverse outcomes at low probabilities of occurrence. He surveyed decisions that he considered to be precedents and identified the nature of the adverse outcomes involved in each case, along with their probability of occurrence, per experts in the original court cases (table 7.1).[27]

Table 7.1   Judge Robinson's Set of Legal Precedents

Cases where courts required disclosure
- 3% chance of death, paralysis, or other injury
- 1% chance of loss of hearing

Cases where courts did not require disclosure
- 1 in 800,000 chance of aplastic anemia
- 1.5% chance of loss of eye
- 1 in 250 to 1 in 500 chance of perforation of esophagus

Judge Robinson used these court decisions in a quasiscientific way to support his argument that "a very small chance of death or serious disablement which dramatically outweighs the potential benefit of the therapy or the detriments of the existing malady may summons discussion with the patient."[28]

Notes

1. *Rolater v. Strain,* Okla. 572; 137 P. 96 (1913).
2. See *Slater v. Baker and Stapleton,* 95 Eng. Rep. 860, 2 Wils. KB 359 (1767); and *Salgo v. Leland Stanford Junior University Board of Trustees,* 154 Cal. App. 2d 560, 317 P. 2d 170 (1957).
3. *Canterbury v. Spence,* 464 F 2d 772 (1972).
4. P. S. Appelbaum, "Rethinking the Conduct of Psychiatric Research," *Archives of General Psychiatry* 54, no. 2 (February 1997): 117–120, quoted from p. 119.
5. See Appelbaum, "Rethinking the Conduct of Psychiatric Research"; P. S. Appelbaum and T. Grisso, "The MacArthur Treatment Competence Study, I: Mental Illness and Competence to Consent to Treatment," *Law and Human Behavior* 19, no. 2 (April 1995): 105–26; and J. W. Berg, P. S. Appelbaum, and T. Grisso, "Constructing Competence: Formulating Standards of Legal Competence to Make Medical Decisions," *Rutgers Law Review* 48 (1996): 345–96.
6. Appelbaum, "Rethinking the Conduct of Psychiatric Research," 118. The study referred to is T. Grisso and P. S. Appelbaum, "The MacArthur Treatment Competence Study, III: Abilities of Patients to Consent to Psychiatric and Medical Treatments," *Law and Human Behavior* 19, no. 2 (April 1995): 149–74.
7. T. A. Posever and T. Chelmow, "Informed Consent for Research in Schizophrenia," *IRB: Ethics and Human Research* 23, no. 1 (January/February 2001): 10–15, quoted from p. 10.
8. Posever and Chalmer, "Informed Consent for Research in Schizophrenia," 10–11.
9. Posever and Chalmer, "Informed Consent for Research in Schizophrenia," 10.
10. Belmont Report, 12–13.
11. Belmont Report, 14.
12. See T. P. Laughren, "The Scientific and Ethical Basis for Placebo-Controlled Trials in Depression and Schizophrenia: An FDA Perspective," *European Psychiatry* 16, no. 7 (November 2001): 418–23; and E. J. Emanuel and F. G. Miller, "The Ethics of Placebo-Controlled Trials: A Middle Ground," *New England Journal of Medicine* 345, no. 12 (September 20, 2001): 915–19.
13. Appelbaum, "Rethinking the Conduct of Psychiatric Research," 119.
14. *Malette v. Shulman* [1990], 67 DLR (4th) 321 (Ont. CA).
15. See E. I. Picard and G. B. Robertson, *Legal Liability of Doctors and Hospitals in Canada,* 3d ed. (Scarborough, Ont.: Carswell Thomson, 1996), 42.
16. *Canterbury v. Spence,* pp. 781–82.
17. S. E. Fienberg, ed., *The Evolving Role of Statistical Assessments as Evidence in the Courts* (New York: Springer-Verlag, 1989), 25.

18. *Reibl v. Hughes* [1980], 2 SCR 885.

19. *Reibl v. Hughes* [1980], 2 SCR 887–88.

20. *Reibl v. Hughes* [1980], 2 SCR 892.

21. *Reibl v. Hughes* [1980], 2 SCR 893.

22. *Reibl v. Hughes* [1980], 2 SCR 894, 925.

23. Picard and Robertson, *Legal Liability*, 84–85.

24. Picard and Robertson, *Legal Liability*, 85.

25. *Reibl v. Hughes* [1980], 114 DLR (3d) 1, 2.

26. *Canterbury v. Spence*, p. 787.

27. The precedent cases are *Bower v. Talmage*, 150 So. 2d 888 (Fla. App. 1963); *Scott v. Wilson*, 396 SW 2d 532 (Tex. Civ. App.1995), aff'd 412 SW 2d 299 (Tex. 1967); *Stottlemire v. Cawood*, 213 F. Supp. 897, 898 (DDC); *Yeates v. Harms*, 193 Kan. 320, 393 P. 2d 982, 991 (1964); rehearing, 194 Kan. 675, 401 P. 2d 659 (1965); and *Starnes v. Taylor*, 272 NC 386, 158 SE 2d 339, 344 (1968).

28. *Canterbury v. Spence*, p. 788.

8

Expanded Senses of Information by Ethicists and a Psychiatrist

J AY KATZ MAKES SEVERAL POINTS relevant to Judge C. J. Laskin's opinion in *Reibl*. Katz makes the distinction between a patient's trust in his or her physician and *blind trust*. According to Katz, trust is achieved through conversations between physicians and their patients, whereas blind trust is the antithesis of conversation. Katz, a psychiatrist and psychoanalyst, was one of the first U.S. commentators and reviewers of the *Canterbury* decision who viewed Judge Robinson's approach as problematic.

Yet Katz also views decisions involving the reasonable person standard as falling short of the goal of conversations in the patient–physician relationship.

> [M]ere disclosure does little to expand opportunities for meaningful conversation, particularly in surrender-prone medical settings, unless patients are also seen as potential participants in medical decisions affecting their lives.... To accomplish that would have required, prior to a promulgation of an informed consent doctrine, an exploration of the complex care-taking and being-taken-care-of transactions which take place between physicians and their patients as well as the tremendous uncertainties inherent in the art and science of medicine. Decision making in medicine ought to be a joint undertaking and depends much more on the nature and quality of the entire give-and-take process and not on whether a particular disclosure has or has not been made.[1]

Katz's view is that "judges' intention behind the doctrine of informed consent was to give patients a greater voice in decision making, to improve the climate of conversation between physicians and patients.... [T]he courts largely undercut this intention ... with profound reservations that revealed their distrust

of patients' capacities to make their own decisions."[2] While Katz emphasizes
the need for a restructuring of the patient–physician relationship, he believes
that the courts have an overwhelming concern for the "patients' capacity for
'rational' decision making."[3]

Katz believed that judges were caught between two issues. In the first place,

> judicial pronouncements calling for the disclosure of facts confronted courts
> with the staggering assignment of specifying what these facts were. . . . Moreover,
> courts' all too single-minded emphasis on risk disclosure made the objective of
> giving patients a greater voice in medical decision making well-nigh unattain-
> able. . . . Treatment decisions are extremely complex and require a more sus-
> tained dialogue, one in which patients are viewed as participants in medical de-
> cisions affecting their lives.[4]

For Katz, "decision-making in medicine ought to be a joint undertaking and
depends much more on the nature and quality of the entire give-and-take
process and not on whether a particular disclosure has or has not been
made."[5]

Katz is also concerned with the courts' sensed need "to promulgate an informed consent doctrine which articulates the extent of communication required for *all* medical encounters." Katz offers a less dogmatic approach:

> For analytic purposes it may be more profitable, at least to begin with, to give
> separate consideration, for example, to the diagnostic, prognostic, and thera-
> peutic facets of medical practice, to acutely and chronically ill patients, to condi-
> tions that can be treated by a variety of means or not at all, and to intervention
> in which faith in the therapy makes a significant contribution to cure.

For Katz, "such an analysis may even reveal that at times compelling reasons
exist for not communicating disturbing information to patients."[6] This perspective has received preliminary study by other investigators.

As Cassell, Leon, and Kaufman observe, "Clinicians have long known that
sick persons, although appearing to have normal mental capacity, may have
difficulty thinking clearly when presented with complex clinical choices."[7] The
authors recruited sixty-three patients and twenty-eight controls. Their study
patients had been consecutively admitted to the thoracic surgery and general
medical services of the New York Presbyterian Hospital in New York City.
Their controls came from a nonresidential senior citizen center. The authors
used seven tasks designed by the psychologist Jean Piaget to assess judgment
in childhood cognitive development. These tasks were

- conservation of volume;
- conservation of substance;

- conservation of length;
- preservation of the horizon;
- conservation of area;
- ability to decenter;
- ability to classify.

They also measured the degree of sickness of a patient by using the Karnofsky scale of physical function.

The authors found that patients with Karnofsky scores of 50 or less responded correctly to fewer tasks than controls ($p < .001$). In addition, a smaller proportion of sicker patients responded correctly to each of the seven tasks. However, patients with Karnofsky scores greater than 50 did not perform differently than controls. The authors conclude that in sicker hospitalized patients, performance on seven tasks of judgment was similar to that among children younger than ten years of age.

> These Piagetian tasks and the mental abilities that they evaluate are mastered during middle childhood (approximately 6 to 10 years of age). The thinking of persons who cannot do these tasks correctly has been described as being focused on particular states; such persons are therefore unable to take into account transitions. . . . Persons who think like this can attend to only a limited amount of information at any one time.[8]

Thus, patients in a state of sickness cannot be assumed to be the competent functioning adults they are when in a state of good health. The authors argue that additional research is needed to evaluate the abilities of sick patients to make sound judgments about "clinical decisions, informed consent, and the execution of legal documents."[9] Although this research needs to be explored from a number of directions, it is important to assess the information message that is given to and must be received by the patient in sickness and in health.

To summarize, the judicial doctrine of informed consent provides general guidance on the types of information that physicians must disclose to patients who are about to embark on a medical procedure. For the courts, in the majority of cases, the main issue in informed consent in clinical care involving competent adult patients is one of risk disclosure as the primary obligation of physicians. To some extent at least, the fact that courts review cases of alleged lack of informed consent after injuries have occurred has led them into this risk disclosure perspective: Had the risk been disclosed, the patient would not have agreed to the intervention. The framework of information proposed by Judge Robinson in *Canterbury* for disclosure of information to competent adult patients—based on risks, benefits, alternatives, and timing of medical interventions—will not necessarily work with a patient who is too sick or too impaired.

The issues of decision-making competence can be seen in a variety of settings. Let us take depression. Carl Elliott notes that "most accounts of competence focus on intellectual capacity and abilities to reason, and depression is primarily a disorder of mood." Yet Elliott wants

> to challenge this account of competence and argue that depression may well impair a patient's competence to consent to research. Most crucially, it can impair a person's ability to evaluate risks and benefits. To put the matter simply, if a person is depressed, he or she may be *aware* that a protocol carries risk, but simply not *care* about those risks. This sort of intellectual impairment can be as important a part of patient competence as the more detached, intellectual understanding that most accounts of competence emphasize.[10]

And so the weighing of risk and benefit becomes a crucial component of decision making in both clinical care and clinical research. We will now move from the judicial stance on information, with its emphasis on risk disclosure, to ethical perspectives that also value scientific information related to medical risk.

Notes

1. J. Katz, "Informed Consent: A Fairy Tale? Law's Vision," *University of Pittsburgh Law Review* 39, no. 2 (Winter 1977): 137–74, quoted from pp. 172–73.

2. J. Katz, *The Silent World of Doctor and Patient* (New York: Free Press, 1984), xvi.

3. Katz, "Informed Consent," 172.

4. Katz, *Silent World*, 82–83.

5. Katz, "Informed Consent," 173.

6. Katz, "Informed Consent," 173.

7. E. J. Cassell, A. C. Leon, and S. G. Kaufman, "Preliminary Evidence of Impaired Thinking in Sick Patients," *Annals of Internal Medicine* 134, no. 12 (June 19, 2001): 1120–23.

8. Cassell, Leon, and Kaufman, "Preliminary Evidence," 1122.

9. Cassell, Leon, and Kaufman, "Preliminary Evidence," 1122.

10. C. Elliott, "Caring About Risks: Are Severely Depressed Patients Competent to Consent to Research?" *Archives of General Psychiatry* 54, no. 2 (February 1997): 113–14.

9

Bioethicists Respond to the Judicial Doctrine of Informed Consent

Certain bioethicists support Judge Robinson's view that risk disclosure should remain a key element of patient decision making. For example, Gert, Culver, and Clouser note that "rationality [itself] is very intimately related to harms and benefits."[1] Yet they use the *irrationality* as their base concept.

> To act irrationally is to act in a way that one knows (justifiably believes), or should know, will significantly increase the probability that oneself, or those one cares for, will suffer death, pain, disability, loss of freedom, or loss of pleasure; and one does not have an adequate reason for so acting.

For the authors, "any intentional action that is not irrational is rational." However, while there may be a complete agreement on what the basic harms are, there may be "considerable disagreement on the ranking of these harms."[2] The authors recognize variation in degrees of harm, in terms of its severity, its occurrence in individuals of different ages, and the intensity with which harms may be felt by different individuals.

However, the authors do not elaborate on one area of interest: the fact that the probability of a harm occurring may not only be different in different individuals but may be understood (or misunderstood) by different individuals in very different ways. Thus, not only is there uncertainty as to when a particular adverse event will occur in a particular patient, there is also some level of uncertainty about how the individual will react to the fact that a risk will occur. This notion is captured by the court's concept of *therapeutic privilege of nondisclosure of information in the case of certain patients.* Judge Robinson held in *Canterbury* that

> patients occasionally become so ill or emotionally distraught on disclosure as to foreclose a rational decision, or complicate or hinder the treatment, or perhaps even pose psychological damage to the patient. Where that is so, the [court] cases have generally held that the physician is armed with a privilege to keep the information from the patient.

But "the critical inquiry is whether the patient responded to a sound medical judgment, . . . for otherwise [the concept of *therapeutic privilege* in which a physician may withhold information in a particular circumstance] might devour the disclosure rule itself."[3] One of the potential uses of written informed consent forms is that, if the patient prefers not to be told about the risks, the physician can ask if the patient would like a family member or significant other to have a copy of the form.

Gert, Culver, and Clouser argue that while "a physician is morally required to provide a patient with adequate information. . . all the 'adequate information' requirement amounts to is reminding physicians not to deceive." The authors argue that in medical practice, the kind of information that passes the "adequacy" test is "that information that any rational person would want to know before making a decision."[4] The authors make the following points:

- There are many clinical situations in which patients face only one rational choice.
- Each rational choice brings with it a particular estimated probability of suffering some harms and of preventing or relieving others.
- Rational persons vary among themselves with regard to how they rank the harms associated with different rational choices, and different individuals therefore make different rational choices.
- It would be unreasonable for a person not to make the choice that reflects most accurately his or her own ranking of the harms associated with the different alternatives.

For Gert, Culver, and Clouser,

> these considerations make it important for rational persons to have full relevant information about the likely harms associated with all existing rational treatment choices. . . . [T]o have less than full information introduces the possibility that patients will make some choice other than the one closest to *their own ranking of harms,* either because they do not even know a choice closer to their ranking exists, or they do not know sufficient detail about the harms associated with the choices of which they are aware.[5]

Thus "the information that rational persons want to know, at a minimum, consists of the probable significant harms and benefits associated with a suggested treatment, with any rational alternative treatments, and with no treatment at all."[6]

The authors provide a list of "basic harms," including death, pain (psychological as well as physical), mental and physical disabilities, and loss of freedom or pleasure. They also list nausea, weakness, and hair loss, and their estimated chance of occurrence, their level of severity, and how long the symptoms are expected to last. Then they list what "patients almost never care about," including the chemical structure of a drug and the drug's postulated mechanism of action. While "less serious harms of a low probability of occurrence often need not be told, . . . any serious harm that is likely to occur should be." The authors believe that "information about risks should usually be given in numerical form, rather than using terms and explanations like 'common,' 'rare,' or 'it hardly ever happens.'"[7] Gert, Culver, and Clouser's approach rests on two assumptions. The first is that "a reasonably precise and useful body of knowledge exists about the comparative harms and benefits of alternative treatments in particular medical situations"; the second is that "individual physicians are knowledgeable about this data-base and therefore able to present a relevant portion of it to their patients."[8]

Gert, Culver, and Clouser's framework can in fact be tested. We will now review research regarding two issues that are needed to answer the real-life questions they posed. First, we will review research on physicians' reports of their willingness to search for information. Second, we will review patients' reports of what they want from their physicians in terms of information, the choices they want to make, and the level of participation in actual decision making they want to have.

S. Barton has examined British physicians' attitudes toward the search for information. Most providers report that they want to base their practice on evidence and that they feel that this evidence will improve patient care. But despite these beliefs, the idea that each health professional should himself or herself formulate questions; search, appraise, and summarize the literature regarding the care of their patients; and then apply that evidence to their care "has proved too difficult alongside the competing demands of clinical practice."[9] Barton asserts that "over 90% of British general practitioners believe that learning evidence-handling skills is not a priority, and even when resources are available, doctors rarely search for evidence." However, "72% do often use evidence-based summaries generated by others, which can be accessed by busy clinicians in seconds."[10]

While educational outreach visits and opinion leaders improve the adoption of evidence by clinicians, printed materials remain ineffective. Although no study has explored this issue in detail, Barton speculates about why "passive distribution of printed materials does not automatically change behavior": "information may have been difficult to access when it was needed, may have been difficult to understand, or may have been irrelevant." Conversely, "printed materials may have lacked credibility without a method of checking that the information is rigorous and complete."[11]

Barton brings up the issue of what constitutes *credible information* based on scientific evidence, and details strategies that target a physician's clinical behaviors to help get evidence into practice. These strategies include "discussions with an expert, academic detailing, advice from opinion leaders, targeted audit and feedback, computerized alerts or reminders, and local development of evidence based policies." Moreover, "combined approaches are more effective [at changing physician behavior] than individual techniques used alone."[12]

Barton argues that a physician's clinical decision has at least three components. First, there is the available scientific evidence (and the further recognition of the fact that in a great many cases the scientific evidence will be poor). Second, there is the physician's clinical expertise in evaluating each patient's circumstances. Third, there is the physician's expertise in evaluating each patient's personal preferences. Barton suggests a set of questions that illustrate a physician's clinical expertise:

- Is my patient typical of those in the studies?
- Are the interventions likely to be delivered in the same way as in the trials?
- Are the reported outcomes for benefits and harms the ones we want to know about?

For Barton, "clinical judgment is then required to estimate the relevance of the best available evidence for an individual and to explore its meaning for them."[13]

Yet the very first of Barton's questions—Is my patient typical of those in the studies? —may be hard for nonscientists to make operational in their clinical practice setting. And his final question—Are the reported outcomes for benefits and harms the ones we want to know about?— depends on who "we" are. We have already said that much of the important quality-of-life information that patients may want to know related to two different management strategies may not have been systematically examined in a given research study because of the costs of collecting quality-of-life data in addition to survival/mortality data.

It is not yet clear what these physicians are doing with evidence-based summaries. One would suspect that physicians are using these summaries to refine their own deliberations and thinking processes regarding clinical questions. However, further research is needed to learn about the extent to which health professionals go in sharing this information with their patients. Such sharing of evidence between a physician and a patient is particularly difficult because it is estimated that only 15 percent of clinical treatment is in fact supported by scientific evidence.[14] What is the remainder of clinical treatment supported by? The answer is, on the basis of the clinical experience of physicians.

Notes

1. B. Gert, C. M. Culver, and K. D. Clouser, *Bioethics: A Return to Fundamentals* (New York: Oxford University Press, 1997), 26–27.

2. Gert, Culver, and Clouser, *Bioethics*, 26–27.

3. *Canterbury v. Spence*, 464 F 2d 772 (1972), 789.

4. Gert, Culver, and Clouser, *Bioethics*, 152–53.

5. Gert, Culver, and Clouser, *Bioethics*, 153–55.

6. Gert, Culver, and Clouser, *Bioethics*, 162.

7. Gert, Culver, and Clouser, *Bioethics*, 163.

8. Gert, Culver, and Clouser, *Bioethics*, 169.

9. S. Barton, "Using Clinical Evidence: Having the Evidence in Your Hand Is Just a Start—But a Good One," *British Medical Journal* 322 (March 3, 2001): 503–04, quoted from p. 503.

10. Barton, "Using Clinical Evidence," 503.

11. Barton, "Using Clinical Evidence," 503–04.

12. Barton, "Using Clinical Evidence," 503–04.

13. Barton, "Using Clinical Evidence," 503.

14. J. Katz, "Reflections on Informed Consent: 40 Years After Its Birth," *Journal of the American College of Surgeons* 186, no. 4 (April 1998): 466–74, quoted from p. 469.

10

Information, Cultures, and Caution

Now we go back to Judge Robinson's view of the medical justification for withholding information from those few patients who can be harmed medically by the information. This is not a new perspective. Some cultures prefer that certain information not be given to patients with certain diseases. This withholding of information most typically occurs in relation to cancers likely to considerably shorten the life span of the patient. But is it not better to ask patients how much information they want, rather than just assuming that they do not want information? Should we provide the information to someone besides the patient?[1]

Prognostic information about diseases and medical conditions like cancer or congestive heart failure poses a difficult dilemma. Weeks et al. studied 917 patients with metastatic colon cancer or advanced non-small-cell lung cancer and found that patients who had optimistic perceptions of their prognosis more often requested invasive therapies than supportive care.[2] The authors, however, did not inquire about the evidence or reports their patients used to formulate their opinions about their prognosis. Elizabeth Lamont and Nicholas Christakis studied 311 patients with cancer.[3] For 300 of them (96.5 percent), physicians were able to formulate prognoses. The physicians reported that

- they would not communicate any survival estimate 22.7 percent of the time;
- they would communicate the same survival estimate they formulated 37 percent of the time;

• they would communicate a survival estimate different from the one they formulated 40.3 percent of the time.

Furthermore, "of the discrepant survival estimates, most (70.2%) were optimistically discrepant." The authors speculate that even experienced physicians recognize that they "cannot formulate reliable prognoses" and therefore do not try to.

> Clearly, however, communication of bad news needs to be handled tactfully and respectfully. . . . Our study should not be taken to support the deplorable practice of "truth dumping." Rather, we believe that physicians need to face the difficulties involved when seriously ill patients insist on temporally specific prognoses. Armed with such information, patients might be better able to plan for, and achieve, the kind of "good death" most Americans say they want.[4]

Truth-dumping is the dumping of information on patients who do not want it, are not ready to receive it, or would prefer that others receive it for them. Truth-dumping can be avoided simply by asking patients how much they want to know. But simply asking the question may beget another: "What is my prognosis, doc?" How good is the physician as an estimator of prognosis?

Peter Ubel argues that nonspecific communication approaches are often sufficient.

> I might say, "I cannot predict the future, but in my experience, patients with your illness typically live a matter of months, not years," or "Many people in your condition will live for only a matter of weeks, but some live significantly longer. I do not know what your fate will be." In these conversations, I discuss concrete treatment goals with patients. I do not hesitate to say when I think the goal should shift from cure to palliation. When things are grim, I suggest that it is time to visit with friends and family because "it is better to be safe than sorry." I give enough prognostic information to help patients make decisions, but I avoid using numerical wording that suggests I have a prognostic crystal ball.[5]

While Ubel's comments here are not completely developed, he agrees with Katz in terms of how physicians should approach patients in general, that is, through conversations. Ubel also reflects what Katz believes these conversations should be about: uncertainty. Yet Ubel advocates a use of *relatively nonspecific prognoses* that reflect in many cases the actual skills of physicians in making predictions in the cases of individual patients.

Another necessary contrast is that between information and security. David Meddings has argued that human security is a prerequisite for health.[6] When an individual's security is destroyed by continual fear and the danger of a death he or she cannot understand or imagine, whether due to disease or war, the psychological impact is tremendous. Thus, information and how it is portrayed in nonspecific terms may still have a tremendous impact on the remaining life of that individual.

Frequently the patient insists on prognoses of specific types, with numbers and their chance of being wrong. Other patients can be approached with nonspecific questions that raise the issue of information and decision making: How much do you want to know about what I'm thinking as a physician and about how I might be wrong in any estimates I may be able to formulate in your case? Or to whom do you want me to speak on your behalf about these issues?

What is needed here is the development of a methodologic framework that clinicians can use to develop their own questions based on the preceding factors. One suggestion is *preemptive conversations* introduced by the physician during health and prior to the onset of disease. Remember, patients can be approached in primary care sectors, general medicine clinics, and other outpatient clinical environments when they are not sick and where more information is available. What have researchers learned about patients' interest in *risk disclosure* or about what rational individuals want to learn concerning various kinds of harms they may face? Researchers in two different areas of medical intervention—medications and invasive procedures—have found evidence that patients agree with Judge Robinson's views that patients want to know about low probability events. For some, risk disclosure by itself may never be enough for decision making. Some will require a keen review and analysis of the nature of the medical risk, the nature of the scientific evidence behind the decision, and the sufficiency of that scientific evidence for decision making.

Let us ask: What information should you present to a patient in clinical care or an individual considering enrolling in a research trial? I frequently hear the following from researchers: "What you have to do is to give the patient 'enough' information." The questions remain: What is "enough" information, and who is to decide this? Once one begins to address these questions, they beget others. How much information is enough information for the *scientist* and for the *nonscientist*? How much evidence is enough evidence? How much scientific evidence is enough scientific evidence? Enough for what?

What individuals would like to hear is that all information, in some way or another, is backed up by solid evidence. The problem is that there are many approaches to viewing what evidence is in fact solid and what is not solid. Indeed, some information may not be backed by any evidence, but only by guesses or estimates. In some cases, these guesses and estimates may be those of national or international experts or even consensus panels of experts. Even in the best of cases, where there is scientific evidence supporting information, there is usually a question about the *accuracy* of the information at the time it is given or obtained and about how *reliable* the information will be over time. We will view scientific evidence as ranging from old to new science. And we will view scientific experts as ranging in terms of their quality and reliability regarding the information they use, the individual-vs.-societal perspective they take regarding resources, and the interpretations of data they provide. Science is not simply for scientists. The scientific expert has a solid place not only within science but also as a spokesperson to the public and as an expert in the courtroom. Scientific evidence has had a long history of development, from Sir Francis Bacon's scientific method through Carl Gustav Hempel's view, to the more recent view of Karl Popper. Yet, scientific theory continues to change, opening up questions regarding how it should be communicated— not only in the physical sciences but also in the health and medical sciences.

It is hard to say how Gert, Culver, and Clouser mean to specify *harms* in their approach. They write, "the basic harms which concern rational persons are: death, pain (psychological as well as physical), disabilities (both mental and physical), and loss of freedom or pleasure." The authors make a number of assumptions about what patients "almost never care about," such as

> the chemical structure of a chemotherapeutic agent, or . . . its postulated mechanism of interference with the metabolism of neoplastic cells. They do care about pain, nausea, weakness, and hair loss—about how likely these are to occur, how severe they are apt to be, and how long they may be expected to last. . . . Unless well-informed patients specifically request technical detail, they [the technical details] should not be given.[7]

What we have in Gert, Culver, and Clouser's specification of basic harms is a large amount of information about one medical intervention—the physician-recommended intervention—but none about the harms (and benefits) of the alternative reasonable interventions. Discussion of alternative strategies needs to include harms and benefits of the nonintervention strategy ("let nature takes it course") and the delay strategy (delay decision making to search for new information, to wait for new research results, and so on). The question remains: How is the information to be made useful to the competent patient? Gert, Culver, and Clouser's approach to harms and their disclosures get us directly to the issue of quantity and quality of information disclosed or communicated.

Quantity of Information Communicated

Judge Spottswood Robinson in the *Canterbury* decision rejected the notion of full disclosure in informed consent:

It seems obviously prohibitive and unrealistic to expect physicians to discuss with their patients every risk of proposed treatment—no matter how small or remote—and generally unnecessary from the patient's viewpoint as well. Indeed, the [court] cases speaking in term of "full" disclosure appear to envision something less than total disclosure, leaving unanswered the question of just how much.[8]

Judge Robinson instead developed the notion of reasonable disclosure:

Optimally for the patient, exposure [disclosure] of a risk would be mandatory whenever the patient would deem it significant to his or her decision, either singly or in combination with other risks. This requirement, however, would summon the physician to second-guess the patient, whose ideas on materiality could hardly be known to the physician. That would make an undue demand upon medical practitioners, whose conduct, like that of others, is to be measured in terms of reasonableness. . . .

The scope of the [disclosure] standard is not subjective as to either the physician or the patient; it remains objective with due regard for the patient's informational needs and with suitable leeway for the physician's situation. In broad outline, we agree that "[a] risk is thus material when a reasonable person, in what the physician knows or should know to be the patient's position, would be likely to attach significance to the risk or cluster of risks in deciding whether or not to forego the proposed therapy."

But he focused not only on disclosure of information, but on disclosure of information about harms:

The factors contributing significantly to the dangerousness of a medical technique are, of course, the incidence of injury and the degree of harm threatened. A very small chance of death or serious disablement may well be significant; a potential disability which dramatically outweighs the potential benefit of the therapy or the detriments of the existing malady may summons discussion with the patient.[9]

Quality of Information Communicated

Annete O'Connor and Adrian Edwards argue that decision aids may have an important role in promoting evidence-based patient choice in medicine and health care.[10] Building on this work, Holmes-Rovner, Llewellyn-Thomas, and Elwyn outline the following plan for such a decision aid:

- display of the patient choice problem for the patient;
- presentation of "all reasonable options" (within policy constraints), including doing nothing and watchful waiting;

- credible, evidence-based statements about benefits and harms associated with each option;
- description of the quality and consistency of the underlying empirical studies;
- uncertainties to be made explicit;
- controversies to be made explicit;
- value trade-offs to be made explicit.[11]

However, in certain instances the full range of information from which information subsets are to be drawn may be incomplete. This incompleteness is related to the medical literature, which is the obvious source of information for quality disclosure or communication. For example, Scott Plous argues that

> academic journals prefer to publish research findings that run counter to ordinary intuitions, and, as a consequence, more research has been published on failures in decision making than in successes. In this respect, professional journals are similar to newspapers and magazines: They place a premium on news that is surprising and intriguing.[12]

Thus, where the information subsets come from is still dependent on changes in the editorial policies of the peer-reviewed medical scientific literature.[13]

In a similar vein, Jan P. Vandenbroucke argues that "the shaping of medical knowledge is . . . the relative role of fact versus opinion, or of empiricism versus theory" and that "general medical journals have been instrumental in making medicine and medical research accept medical ethics and medical ethics committees, if only by stating that they would no longer publish experiments on human beings unless they had been approved by an ethics committee."[14]

In addition to Plous's noting the influences on what gets published in the peer-reviewed literature and Vandenbroucke's view of the positive impact of this literature in the direction taken by clinical care research, there is room for further discussion of improvements. Saint et al. surveyed 399 eligible physicians in internal medicine across the United States. Of these internists, 264 (66 percent) returned the surveys. Their respondents reported "spending 4.4 hours per week reading medical journal articles. . . . These findings were similar for internists with and without epidemiology training." The study suggests that "peer review and pre-screening of articles by journal editors are highly valued by clinical readers and thus should remain an essential component of clinical journals."[15]

Frank Davidson, former editor of the *Annals of Internal Medicine*, argues that "the efficacy of peer review in guiding publishing decisions and improving the editorial quality of published manuscripts is now being intensely studied. This is as it should be because it's an instrument of science and, like any such instrument, it can and should be improved as needed."[16]

Even when quality information is available for presentation, the question arises as to how it is understood. Plous points to problems like the primacy and recency effects:

> [A] primacy effect sometimes occurs when people are exposed to opposite sides of a controversial issue. In many cases, people are more influenced by the first presentation of an issue than by subsequent presentations. . . . This is not always the case, however. In some instances, the final presentation has more influence than the first presentation. Such a pattern is known as the "recency effect."[17]

Harms need to be associated with real-world descriptions in terms of survival and quality of life, and of real probabilities of occurrence. Yet there may be many problems regarding how individuals differ in what they consider to be a *viable existence* for themselves.

First, the "quality of life" issues may not have been satisfactorily studied because of lack of interest on the part of the principal investigator or funder. Or the funder or principal investigator may not have made a concerted effort to secure a suitable range of patients with different severities of disease. Patients would like to know how other patients like them fared in a study in terms of their quality of life on a new treatment as opposed to a standard treatment. Yet the criteria of the study may have excluded all complex patients with more than one disease or with multiple medications, for valid scientific reasons.

Second, the patient's disease may be an orphan, like Fabry's disease, which occurs in too few instances to generate concerted interest in or funds for study. Another factor may be that a disease is not easily amenable to study. For example, certain diseases—such as types of localized prostate cancer with low Gleason scores—may be associated with a long life in a patient before morbidities and death from the disease occur. Studies of these types of diseases require long periods of time before results are achieved and can be interpreted.

Ranking Benefits and Harms in Relation to Evidence Available

Therefore, I am suggesting that patient rankings for decision making need at least six dimensions.

1. *Harms.* Harms are understood as adverse outcomes associated with drugs, drug withdrawals, devices and their implantation, and other medical strategies, including nonintervention, intervention at a later time (delayed intervention), and watchful waiting.
2. *Benefits.* Benefits are broadly understood in terms of a basic distinction between *survival* and *quality-of-life* benefits.

3. *Scientific evidence.* Here *evidence* or *scientific evidence* is to be distinguished from *information* or *scientific information.* In general, *evidence* is
something that tends to *prove* or *provide grounds for holding a belief.*[18]
The term *scientific evidence* implies that science is the thing that supplies
the proof or basis for belief. Wolf argues,

> Evidence obtained from at least one good, properly designed randomized
> clinical trial traditionally has been considered the highest quality evidence
> available (e.g., see the U.S. Preventive Services Task Force). Evidence from
> a properly designed systematic review of all the available properly de-
> signed randomized trials has been deemed to be evidence of the highest
> quality for use in decision making, and it is superior to the results from
> any one randomized clinical trial.[19]

4. *Clinical experience.* Clinical experience may involve both international
and national consensus panels of experts as well as the personal physicians caring for individual patients. When there are no randomized clinical trials available and a medical condition has not been studied, these
clinical experiences form the basis of care decisions until such research
studies are begun.
5. *Estimation experience.* Estimation experience focuses on the abilities to
estimate the chance of an event occurrence. While estimation experience
may be related to *understanding of scientific evidence* and *clinical experience,* it may be the least studied of the areas of the art and science of
medicine.
6. *Psychological experience involving estimates.* Psychological experience involving estimates includes the basic frameworks that individual physicians use in making their estimates related to their individual patients. It
also includes the psychological and other value-based reasoning of the
cognitive states of the patient with a medical illness that determines
whether disclosures are made, when disclosures are made, how (in what
formats) disclosures are made, and to whom disclosures are made.

A More Elaborate Framework

Yet there are factors that need to be taken into consideration beyond these six
basic dimensions. Attention needs to be focused on six additional components in any ranking or prioritizing framework.

1. *Use of risk-benefit ratios.* We can see that the complexity of the overall
ranking process and the use of summary measures like the risk-benefit

ratio may well hide key data from a patient. This hiding of information occurs every time a risk-benefit ratio is *reduced to lowest terms*. In such a case, the risk-benefit ratio hides the "true numerator" and "true denominator" from which the reduced ratio was derived. Therefore, the use of risk-benefit ratios reduced to lowest terms should be minimized or eliminated as a serious summary measure in individual ranking and decision making.

2. *The battle between overinforming and underinforming.* There is a constant battle regarding what constitutes overinforming and underinforming. What are the results of overinforming patients? Of underinforming?

3. *The battle between the "setting" of information disclosure and patient–physician conversations.* This point is missed by Katz, who assumes that conversations between patients and physicians are sufficient for informing and ultimately making decisions. Some patients may want to go over information and digest the material in a setting where the physician is not even present.

4. *The battle between "oral discourse" and "printed information."* Some patients may not be able to digest information provided only in oral discourse. They want to have written information available to read through on their own, away from the context of the physician providing information. For example, printed diagrams may better illustrate how a cardiac catheterization procedure is conducted than oral discourse.

5. *The battle over the precise "presentation format" used to inform.* Some patients may prefer to have information provided in formats other than the number of patients surviving a procedure. For example, it is difficult for a patient to understand a five-year survival curve comparing two treatments without presenting the curve. Some patients may just want to see the survival/mortality numbers in a table. Some patients may want to forego consideration of any of the numbers and have their physicians use verbal probability terms. Other patients, even though they may believe it is their legal or ethical right to have the information, may still just place the information in a drawer after receiving it.

6. *Systematic vs. intuitive decision making.* More systematic patients may want to weigh risk-benefit information in more formal and mathematically intense ways. More intuitive decision makers may just listen and formulate an opinion without any additional mathematical structure being placed on their decision problem. For some patients, intuitions may be important as an additional or even the sole basis for decision making. Yet, the specific area of intuitive decision making has been understudied and underdeveloped within medical decision making.

Notes

1. See M. Kuczewski and P. J. McCruden, "Informed Consent: Does It Take a Village? The Problem of Culture and Truth Telling," *Cambridge Quarterly of Healthcare Ethics* 10, no. 1 (Winter 2001): 34–46.

2. J. C. Weeks et al., "Relationship Between Cancer Patients' Prediction of Prognosis and Their Treatment Preferences," *Journal of the American Medical Association* 279, no. 21 (June 3, 1998): 1709–14.

3. E. B. Lamont and N. A. Christakis, "Prognostic Disclosure to Patients with Cancer Near the End of Life," *Annals of Internal Medicine* 134, no. 12 (June 19, 2001): 1096–1105.

4. Lamont and Christakis, "Prognostic Disclosure," 1096.

5. P. A. Ubel, "Truth in the Most Optimistic Way," *Annals of Internal Medicine* 134, no. 12 (June 19, 2001): 1142–43.

6. D. R. Meddings, "Human Security: A Prerequisite for Health," *British Medical Journal* 322 (June 23, 2001): 1553.

7. B. Gert, C. M. Culver, and K. D. Clouser, *Bioethics: A Return to Fundamentals* (New York: Oxford University Press, 1997), 163.

8. *Canterbury v. Spence,* 464 F 2d 772 (1972), 786.

9. *Canterbury v. Spence,* 787–88.

10. A. O'Connor and A. Edwards, "The Role of Decision Aids in Promoting Evidence-Based Patient Choice," in *Evidence-Based Patient Choice: Inevitable or Impossible?* ed. Edwards and G. Elwyn (Oxford: Oxford University Press, 2001), 220–42.

11. M. Holmes-Rovner, H. Llewellyn-Thomas, and G. Elwyn, "Moving to the Mainstream," in *Evidence-Based Patient Choice: Inevitable or Impossible?* ed. A. Edwards and Elwyn (Oxford: Oxford University Press, 2001), 278.

12. S. Plous, *The Psychology of Judgment and Decision Making* (New York: McGraw-Hill, 1993), xv.

13. Plous, *Psychology of Judgment,* 42.

14. J. P. Vandenbroucke, "Medical Journals and the Shaping of Medical Knowledge," *Lancet* 352 (December 19/26, 1998), 2001–06, quoted from pp. 2001 and 2006.

15. S. Saint et al., "Journal Reading Habits of Internists," *Journal of General Internal Medicine* 15, no.12 (December 2000), 881–84, quoted from p. 884.

16. F. Davidson. "Editor's World," *Annals of Internal Medicine* 134, no. 12 (June 19, 2001), 1144–46, quoted from 1146.

17. Plous, *Psychology of Judgment,* 42.

18. See F. M. Wolf, "Summarizing Evidence for Clinical Use," in *Evidence-Based Clinical Practice,* ed. J. P. Geyman, R. A. Deyo, and S. D. Ramsey (Boston: Butterworth-Heinemann, 2000), 134.

19. Wolf, "Summarizing Evidence," 137.

II

How Information Reaches Patients

11

What Has Been Learned in Research Studies about "Information" in the New Medical Conversation?

CONTEMPORARY RESEARCH BY PHYSICIANS and other investigators has focused on particular types of information that are now challenging the patient and the physician in the new medical conversation.

Frequently, research related to "information" is carried out through surveys or structured interviews. Ziegler et al. studied how much information about adverse effects of medication patients want from physicians.[1] The authors surveyed 2,500 sequential adults visiting outpatient clinics; 2,348 (61.2 percent women and 38.8 percent men, mean age 47.2 $\pm$ 16.9) responded. Each filled out a twelve-item questionnaire asking whether they would want to hear of *any adverse effects, no matter how rare,* and whether they wanted to hear of *any serious adverse effects, no matter how rare.*

Regarding *any adverse outcome, no matter how rare,* 76.2 percent of respondents reported that they would want this information, while 83.1 percent reported that they would want to hear *any serious adverse outcome, no matter how rare.* The authors note that "this was the opinion expressed most frequently by all age groups, both sexes, and all educational-level groups." Surprised by these findings, the authors asked, Does this mean that most individuals do not discriminate between the two (*all adverse effects* vs. *serious adverse effects*)? Although patients expressed a desire for all information about risk for adverse effects, no matter how rare, these patient responses seemed "unrealistic" to professionals: "probably, many did not realize the magnitude of the mass of information involved and gave a stereotyped response on the order of 'more knowledge is good.'" The authors also note that "subjects who feel uncertain when dealing with numbers in estimating risk and who prefer qualitative to quantitative terms might also tend to choose this alternative [information about risk for adverse effects, no matter how rare]."[2]

Yet the authors overlook several points. First, patients may be interested in nonserious side effects because they would like a ready source of information to see if any nonserious side effects they experience could be related to the medication. Second, patients may want to peruse a long list of side effects to better understand the medication they are on as a therapeutic agent. Third, patients may want to know all the side effects purely for *quality-of-life* reasons. Many physicians believe that physicians should tailor patients' medications to the individual's quality-of-life concerns.

An example of such medication-tailoring involves hypertensive agents. Carola Bardage and Dag Isacson sent a postal questionnaire to a random sample of 8,000 inhabitants (ages 20 to 84 years) in Uppsala County, Sweden, in 1995.[3] Their response rate was 68 percent. These authors found that nearly 20 percent of the users of antihypertensive drugs reported side effects. Men and women reported side effects to nearly the same extent. Individuals with hypertension, with or without medication, rated lower health utilities (-7.1 and -6.0, respectively) than did normotensive individuals after controlling for age and sex. The lowest value, -8.7, was found among drug users who experienced side effects. The side effects that had the strongest negative impact on health utility included impotence and emotional distress related to insomnia, fatigue, and depression. The authors concluded that side effects among hypertensive patients are common and that both the disease and the drug treatment adversely affect the patient's well-being. However, drug treatment was of less importance than prior studies had found. The findings here suggest that side effects should be taken into greater consideration when evaluating drug treatment.

The patients in Ziegler et al.'s study may have been asking for information regarding how the drug was affecting them so that they could work with their physicians to tailor therapy to the best drug treatment regimen they could achieve without side effects. As Ziegler et al. suggest, many professionals may not be providing this information to their patients in any format, written or electronic, but patients were reporting that they in fact did want the information. Ziegler et al. did not put any cost on the information, and patients were reporting that they would want extensive information at least in those cases where cost of obtaining the information was not an issue. Finally, Ziegler et al.'s patients may have felt that they had an ethical and/or a legal right to the information.

In the case of drug treatment for hypertension, it is clear that at least some authors believe there is an opportunity to tailor the selection of blood pressure medications to a patient's preferences regarding quality of life. Quality of life on antihypertensive agents is a well-studied area, although lacking clear implementation guidelines as to how to tailor medications satisfactorily for individual patients. However, the case of *an individual's genetic information* and its uses, although much less understood in terms of empirical science, is becoming increasingly recognized as an information issue for patients. To the best of my knowledge, the issue of genetic information is perhaps the most recognized issue of a patient's need to be informed because of future consequences. The future consequences of information go well beyond the individual subjects asked to donate specimens of their own cells, blood, tissue, or organs for present or future research. They extend to living relatives and future generations of their family line.

Notes

1. D. K. Ziegler et al., "How Much Information about Adverse Effects of Medication Do Patients Want from Physicians?" *Archives of Internal Medicine* 161, no. 5 (March 12, 2001): 706–13.

2. Ziegler et al., "How Much Information," 711.

3. C. Bardage and D. G. Isacson, "Self-Reported Side-Effects of Antihypertensive Drugs: An Epidemiological Study on Prevalence and Impact on Health-State Utility," *Blood Pressure* 9, no. 6 (2000): 328–34.

12

Complex Risk Information: Genetic Information and Future Generations

THE CONTEMPORARY ISSUES IN GENETIC INFORMATION are twofold: *genetic ownership* and *genetic privacy.* These issues are being fought within Congress and state legislatures today.

Genetic Ownership

Genetic ownership was addressed in the 1990 California State Supreme Court decision, *Moore v. Regents of the University of California.* The majority opinion held that

> [r]esearch on human cells plays a critical role in medical research. This is so because researchers are increasingly able to isolate naturally occurring, medically useful biological substances and to produce useful quantities of such substances through genetic engineering. These efforts are beginning to bear fruit. . . . The extension of conversion [the wrongful exercise of ownership over personal property belonging to another] law into this area will hinder research by restricting access to the necessary raw materials.[1]

The Oregon Privacy Act (Senate Bill 276) was passed in 1995. But additional bills (e.g., Senate Bill 114, January 2001) would weaken the property rights of Oregon's original genetic privacy law.

Genetic Privacy: Genetic Risk Information

No matter how the genetic ownership issues will be settled, genetic privacy will remain of paramount importance. The term *genetic privacy* refers to the *right of privacy that is accorded to individuals regarding the use(s) of their genetic information for research purposes.* The *right to privacy of genetic information* is assumed to be a right of an individual regarding the donation of his or her cells for research purposes, and not a more general right of the blood relatives.

The National Cancer Institute's approach to genetic risk information demonstrates a heightened sense of the seriousness of risk information related to cancer genetics. The institute examines two areas, risk assessment and risk counseling, and recommends that individuals receive genetic counseling before being offered genetic testing. Genetic counseling "gives [the patients] time to fully understand both the various medical uncertainties and psychosocial risks and benefits of this information." The components of this process include "medical, genetic and counseling components such as constructing and evaluating a pedigree; eliciting and evaluating personal and family history; and providing information about genetic risk."[2]

The counseling component includes four phases: pretest counseling, testing, post-test counseling, and follow-up. In pretest counseling, patients should know their rights related to privacy of genetic information and the protections offered in terms of anonymity and confidentiality of data, and of limitations on the protection of such data. Error rates in testing and the uncertainty of the answers that the tests provide should also be discussed. Testing means the test is carried out. Post-test counseling should refer to discussions that take place regarding the actual test results and their interpretation, and limits on the confidence of any interpretation.

Follow-up should concentrate on the emotional, psychological, and intellectual impacts the information may have on the patient over time. The key point here is that information, once divulged, cannot be withdrawn. There are few societal rules that have been constructed regarding how individuals should handle genetic information personally, interpersonally, and within society and its legal structure. Indeed, today there is often little or no legal protection for the individual agreeing to donate cells or tissues for genetic research if such information should get into the hands of an employer or insurer. The release of this genetic information from researchers' secured databases can affect not only the individual donating the cells or tissues but also their blood relatives and future generations.

The National Cancer Institute's focus on the genetic counseling experience

begins by addressing the counselee's needs, worries, questions, and concerns, and giving some indication of what to expect. Early in the session it is essential to establish trust, a sense of safety, and rapport. Other tasks during the early phase of counseling include eliciting beliefs about cancer, understanding questions, and negotiating a mutually agreeable agenda. Later in the session, counseling focuses on obtaining and giving information, promoting autonomy in decision-making, and recommending active coping strategies, while assisting in dealing with practical and other emotional issues. Depending on the circumstances, the counselor may use modeling or role playing, problem-solving, relaxation, reframing, or other techniques.

In summary, "genetic counseling involves some combination of rapport-building and information gathering; establishing or verifying diagnoses; risk assessment and calculation of quantitative occurrence/recurrence risks; education and informed consent processes; psychosocial assessment, support and counseling appropriate to a family's culture and ethnicity; and other relevant background variables."[3]

The decision a patient faces regarding genetic testing for a *specific cancer* includes the following considerations:

- Inherited cancer risk has implications for the entire family.
- Cancer risk counseling may be initiated by an individual but has implications for the entire family, including future generations.
- Misuse of cancer and genetic risk information can have potentially severe implications for the entire family involving insurance and employment status.
- Persistent uncertainty exists and will continue to exist about the risk status of an individual and those genetically related to that individual.
- Uncertainty exists and will continue to exist about the accuracy of testing and the rates of false positives and false negatives.
- Testing is often costly.
- Discussions related to genetic risk information and present and future generations can alter, strain, and disrupt relationships with families.
- Guilt will exist in genetically identified individuals about possible transmission of genetic risk to children.
- If protected information is released (because no protection system is perfect), there is the chance of social discrimination.
- Disclosure of genetic risk information can alter the individual's sense of self, life course, family, and future generations.
- There may be survivor guilt.
- A false-negative result may give an individual *false assurance* that leads to the neglect of continued routine surveillance.
- There may be regret over previous decisions related to the risk of cancer that may be transmitted genetically—for example, prophylactic surgery for breast cancer in women with a strong family history of breast cancer.

• There may be attempts at coercion on the part of family members when families cannot come to an agreement on whether a test should be performed.
• The family may be divided between those who want to be tested and those who do not want family members tested.

The National Cancer Institute presents an array of supportable approaches to informed consent, such as

• including "an informed consent component in counseling";
• "providing opportunities for patients to review their informed consent forms during the genetic testing and counseling process";
• requesting "a second written consent prior to the disclosure to the counselee of genetic test results";
• "providing an opportunity to formally address the issue regarding the counselee's willingness to share genetic test results with interested family members."

The National Cancer Institute's document notes that "obtaining written permission to [share genetic test results with interested family members] can avoid vexing problems in the future should the counselee not be available to release their results."[4]

Here, we have a true sense of heightened awareness related to the potential negative impacts of information on a patient's life and a family, including future generations. One question for future research relates to the appropriate format for an informed consent form. The written informed consent issue is much more complex in the case of genetic information than when we were simply discussing clinical care or clinical research.

Notes

1. *Moore v. Regents of the University of California et al.*, 793 P 2d 479 (Cal. 1990); 271 Cal. Rptr. 146 (1990).

2. National Cancer Institute, "Elements of Cancer Genetics Risk Assessment and Counseling," September 2000, at www.cancer.gov/cancerinfo/pdq/genetics/risk-assessment-and-counseling (accessed August 7, 2002).

3. National Cancer Institute, "Elements of Cancer Genetics Risk Assessment and Counseling."

4. National Cancer Institute, "Elements of Cancer Genetics Risk Assessment and Counseling."

III

Communicating Risk-Benefit
Information Today

13

Truth and Bias in the Way Information Is Presented

W^E NOW MOVE FROM TRUTH as facts and opinions to a quite different area, that of truth and bias regarding the way information is presented. Here, the term *biases* will refer to the cognitive biases. We will consider separately two aspects of choice and decision making: first, the impact of the way information is presented and second, the amount of information that is presented.

1. How the Way Information Is Presented Influences Choice

We are now going to examine studies that manipulate an aspect of the short information message to determine how such manipulations influence individual choice. My colleague and I have asked 436 patients (mean age, 65.2 years; mean level of formal education completed 12.6 years) how they preferred to receive chance information about risks.[1] Of these patients, 42.7 percent preferred disclosure of information about the chance of adverse outcomes using qualitative (verbal) expressions (e.g., *rare, probable, possible*); 35.7 percent preferred disclosure in terms of quantitative (numeric) expressions of probability, such as percentages; and 9.8 percent preferred disclosure in both qualitative and quantitative terms.

We also found that there were wide variations in the numerical equivalent that patients assign to verbal probability terms.[2] When patients and physicians hold discussions about risk and chance of adverse outcomes occurring, they may not recognize that they do not share the same meaning (or range of meaning) when they use verbal probability expressions like *rare, possible,* and *probable.* That is, both patients and physicians may assume that they attach the same or similar meanings to a verbal probability expression like *rare,* where in fact their meanings may be magnitudes off. Our research suggests that patients believe that there is *more shared meaning* between themselves and their physicians regarding the use of *verbal probability terms* than may really exist. In post-interview discussions, patients report that they believe that the numerical equivalent they assign to a verbal probability expression is the same number as anyone else would assign.

Simply pointing out to patients the results of our study did not necessarily help them, because they had worked with the belief of *shared meaning of verbal probability expressions* for such a long time that it was not easy for them to consider using numerical probability expressions, like percentages. Indeed, Woloshin et al. have argued that "limited numeracy may be an important barrier to meaningfully assessing patients' values. . . ."[3] Here, the term *numeracy* refers to an individual patient's ability to use and work with numbers.

Presentation of *short information messages* can influence choice in other ways as well. For instance, *framing effects* can influence how information is described and how data are displayed in graphs, tables, and other formats used to communicate information. These effects can be assumed to be most dominant when information is described in abbreviated information messages, where there is little space to discuss how bias could influence decision making. In their pioneering work, Tversky and Kahneman studied framing in choice situations, asking individuals to choose between two alternatives, one framed positively and the other framed negatively.[4] It is often argued that, from a normative standpoint, *frame* should have no effect on choice.[5] Yet it does.

These small manipulations of information affect choice and decisions in the case of both patients and physicians. These effects are usually demonstrated through surveys and questionnaires where individuals are asked to make choices based on a set of information. Patients are randomly assigned to one of two groups. Each group gets the same information, but in a different frame.

If one is interested in *normative decision making,* that is, how decisions can be optimized in a particular way, the frame should be considered inconsequential. But surprisingly, slight manipulations or framing effects alter the choices of both unsophisticated and sophisticated decision makers.[6] The term *manipulation* has at least two senses here. First, there is manipulation of information. Second, there is manipulation in the sense of an intentional influence on how patients perceive information and how they choose or decide based on the information they receive.

Yet a respondent to a survey or questionnaire may not be responding to the information presented. Few studies actually ask respondents *why* they made the choice they did and whether they based their choice on the information provided or on some other set of information. For example, my colleague and I studied men's reported choices regarding management of asymptomatic localized prostate cancer.[7] We asked men to choose between surgery or watchful waiting based on a set of survival and quality-of-life interventions. When we asked the men how they had made their decisions, some men reported that they based their decision on the experience of others whom they had known personally, men who had undergone surgery, radiation therapy, or watchful waiting. These men reported that they did not base their decisions on any of the scientific information about risk and benefit provided to them in the questionnaire.

The frames used to describe outcomes of interventions, for example, in public health or medicine, are not complex. In fact, they are quite simple. In Tversky and Kahneman's original study, the positive frame was "numbers of lives saved," whereas the negative frame was "number of lives lost." Framing, especially *outcome framing*, as a design issue for information messages, can be interpreted in either a positive or negative light.

Potential Positive Impacts of Framing

We will first focus on the potential positive influences of framing related to getting individuals to change their behavior in a particular way. It must be recognized, however, that this rests on an interpretation of what is positive. In our context, we will consider behaviors that are changed from those carrying risks of morbidity and mortality to those that decrease such risks.

Edwards et al. found the largest effects on patient choice when *relative risk information* is presented. The authors determined that "'loss framing' is more effective in influencing screening uptake behaviors than 'gain framing.'"[8] It can be argued *loss framing* is designed to influence individuals to actually accept and take up screening for a particular disease. It is assumed that the acceptance of a recommendation to be screened for a disease is positive in both an individual and a societal sense. Examples of such screening could include identifying asymptomatic patients with high lipids so that their lipid profile can be managed through various modalities, or encouraging patients to get their blood pressure checked so that hypertension can be found and treated.

In these examples the benefit of screening is well accepted. The question remains whether similar effects would be seen when the tests are less reliable, the disease is less predictable, the disease is less treatable, or the individual does not consider the disease serious.

Potential Negative Impacts of Framing

Framing has potential negative impacts, understood as obstacles that have to be overcome to make an effective or optimal medical choice or decision. Regarding framing influences on informed choice, Edwards et al. note that "when the format of information is altered by presenting more information, and which is more understandable by the patient, this is associated with a greater wariness of the treatments available."[9] However, it must be recognized that such *manipulations* of information influence not only patients but also physicians.[10] In addition, the use of the term *wariness* by Edwards et al. may itself be considered an example of framing. Why did Edwards et al. not use the term *awareness* or *caution*?

In a broad sense, information can be intended in many ways: as education, as information for decision making, as information for respect, and as information for persuasion or promotion. A problem exists when the hearer of the message does not understand how the message is *intended*. Misunderstanding intent of information content is a special danger of the short information message. Let us assume that the information message we are talking about is for educational or decision-making purposes. In this setting, information should be presented in such a way as to make it easily understood. Also, framing effects have the potential to cause individuals to view information in a "more positive" or a "more negative" light than is rational from a decision-making standpoint. Framing effects can be intentional or unintentional. However, it can be argued that since framing effects are known to exist, their use should be minimized in circumstances where decisions are based on information provided by others.

Attribute Framing vs. Goal Framing

Levin, Schneider, and Gaeth make a distinction between *attribute framing*, which affects the evaluation of object or event characteristics, and *goal framing*, which affects the persuasiveness of a communication.[11] Their study involved labeling ground beef as either "75 percent lean" or "25 percent fat." The authors found that a sample of ground beef was rated as better tasting and less greasy when it was labeled in a positive light (75 percent lean) than a negative light (25 percent fat). "The information framed here," they write, "is not the outcome of a risky choice but an *attribute* or characteristic of the ground beef that affects its evaluation." Attributes like "75 percent lean," probability of winning, and chance of surviving frame information positively. Attributes like "25 percent fat," probability of losing, and chance of dying frame information negatively. Levin, Schneider, and Gaeth also argue that "topics involving issues of strongly held attitudes or high personal involvement are less susceptible to attribute framing effects."[12] The authors are quick to note that this finding reveals important aspects of cognitive processing. For example, the valence of a description may have a substantial influence on the processing of that information. Perhaps, positive *and* negative attribute framing should be tested in terms of impact on choice and decision making.

In *goal framing*, the authors note that "the impact of a persuasive message has been shown to depend on whether the messages stress either the positive consequences of performing an act or the negative consequences of not performing the act." The authors point to Meyerowitz and Chaiken's research showing that "women were more apt to engage in breast self-examination when presented with information stressing the negative consequences of not engaging in breast self-examination."[13]

Levin, Schneider, and Gaeth make three key points. First, "goal framing effects—despite their unique features—like other framing effects, may be enhanced, eliminated, or even reversed by a variety of characteristics of the situation." Second, framing can produce both cognitive and motivational consequences. Third, a valence (positive/negative) account of framing should "encourage us to think about the broad-based underlying motivations that guide our day-to-day behavior."[14] The point as regards the patient–physician relationship or human subject–principal investigator relationship is that even when information is suitably framed in a written informed consent form, the physician or principal investigator can influence patient opinion concerning therapy or participation in a research project by subtle uses of framing in oral discourse.

Should framing effects be a surprise? Frames are usually specified as positive or negative (saving lives vs. lives lost; survival vs. mortality). Levin, Schneider, and Gaeth suggest that the human mind may code information in these terms for quick decision making. If such coding is an integral part of human cognitive processing, then the need to understand such influences is key to all choice and decision-making circumstances.

2. The Amount of Information Presented for Choice

We have so far directed our attention to the examination of the way information can be framed and presented, and how that frame or presentation can be manipulated. We now turn our attention to the examination of the amount or completeness of the information provided to patients. Two points are highly relevant to risk (or risk and benefit) information. First, the description of the risk (or risk and benefit) can be captured in a short phrase that is understood by the patient or the human subject. Second, its chance of occurring can be expressed verbally in terms such as *rare, possible,* or *probable,* or expressed numerically in terms such as percents.

Several researchers have examined the impact of the amount of information presented on individual choice or decision making. In the examples we consider next, it will be important for the reader to ask himself or herself whether a thorough listing of three serious potential outcomes is more or less complete than a less thorough listing of twenty very rare yet mild outcomes. Also, is completeness really just a statement of the quality of information? Some would argue not. Let us see what the scientific investigators say regarding the completeness of information messages about *risk.*

Joel Davis studied incomplete risk statements (as provided by drug manufacturers) and complete risk statements. However, it is important to recognize that Davis's complete risk statements included only "all side effects occurring at an incidence of 3% or more" and these side effects, in turn, "were identified from data provided in the brief summary that accompanies each drug's DTC [direct-to-consumer] advertisement."[15]

In Study 1, Davis first conducted individual interviews with undergraduate and graduate students at a major western university. These students were between 22 and 43 years of age and reported that they were concerned with an incidence level of 3 percent or greater as their baseline of concern. Therefore, Davis used complete risk statements but only for risks occurring at a rate of ≥3 percent. However, no numeric incidence levels were provided in the risk statements. Patients received only the name of the side effect.

Davis mailed 140 adults by mall intercept. Respondents who agreed to participate were taken to a small interview room and asked to complete a survey. Of these 140 adults, 75 responded to the questionnaire containing incomplete risk statements, and 65 responded to the questionnaire with complete risk statements. Davis found that "consumers were more likely to 'recommend or purchase' the drug when the drug's description was accompanied by an incomplete versus complete risk statement." However, Davis also found that

> risk statement completeness . . . did not affect *directionality* of response as all mean ratings were greater than the neutral midpoint of the scale; that is, all mean ratings indicated a positive likelihood to recommend or purchase. Thus, regardless of risk statement completeness, respondents indicated that they were more rather than less likely to purchase or recommend a described drug.[16]

In Study 2, complete risk statements included "all side effects with an incidence greater than or equal to the side effect with the lowest incidence mentioned in the advertiser-created risk statement." Moreover, numeric risk incidence levels were provided. The population in Study 2 consisted of 58 adults recruited via mall intercept and was conducted similarly to Study 1. Respondents were asked to rate the safety of the drug on a scale of 0–9 (0 = not safe at all; 9 = extremely safe). Davis found that "drugs described with incomplete descriptions of side effects always were rated as more safe."[17] This area of research is potentially important because it begins to analyze the potential influences of partial information on individual responses to the safety of medication.

Davis's studies do not speak to an individual's mindset when he or she receives a short information message in general. Bell, Kravitz, and Wilkes found that about half of consumers believe that direct-to-consumer ads must receive government approval prior to display, and 43% of consumers believed that only "completely safe" drugs are allowed to be advertised.[18] Davis notes that "consumers' mindset with regard to advertised drugs may be a contributing factor to the pattern of response" to short advertised messages by drug manufacturers. Davis continues, "consumers [react] positively to all advertised drugs . . . because they believe that a drug must be 'good' if it is allowed to be advertised."[19] Perhaps informational messages should contain warnings identifying and clarifying these common misperceptions.

Davis's study also illustrates the way the choice of threshold probability impacts the amount (or completeness) for choice. Davis used a numeric probability threshold in both studies. However, the use of probability thresholds like 2 percent or 3 percent cannot be considered in any sense a complete description of the risks. They may not capture a drug's most severe adverse outcomes (e.g., death and stroke) if they occur less than 1 percent of the time. It is clear from other studies that patients want to be informed of severe adverse outcomes at well below the 1 percent level.

An example of short information messages can be seen in the attempt by the FDA to provide a low level of guidance to drug manufacturers on how to advertise prescription drugs to consumers. Although there are many medical concerns and ethical issues related to drug advertising,[20] the FDA clearly is directing its guidance to the development of drug advertisements, and not primarily to patient education. Although the medical and ethical concerns are relevant, we should first examine the FDA's baseline guidance to drug manufacturers.

The FDA requires a "fair balance" in the text of a drug advertisement. According to the FDA, a product-specific advertisement

> fails to present a fair balance between information relating to side effects and contraindications and information relating to effectiveness of the drug in that the information relating to effectiveness of the drug is presented in greater scope, depth or detail . . . and their information is not fairly balanced by a presentation of a summary of true information relating to side effects and contraindications of the drug.[21]

What needs further clarification is what a failure looks like in particular circumstances. The question regarding the FDA's approach is whether this *fair balance* of risks and benefits has meaning to patients in the ways they prefer to use information.

Little research to date has studied what types of information impact patient decision making. Judge Robinson focused on disclosure of severe adverse outcomes at low chance of occurrence in the creation of his reasonable person approach to informed consent. In addition, there is speculation that once an individual sustains a severe adverse outcome related to a drug, his or her perspective on the importance of future disclosures about severe adverse outcomes at low chance of occurrence will be changed.

There are several ways of altering information to demonstrate framing influences. One way is to provide patients information at lower and lower frequency thresholds, to see how their choices change. For example, does a patient make a similar decision when a chance of an upset stomach is 1 in 1,000,000 as when that chance is 1 in 1,000? However, simply lowering the probability threshold misses an important aspect of information related to choice. That is, the most severe adverse outcomes in medical care often have lower probabilities of occurrence. Thus, the severity of an illness may be inversely proportional to its chance of occurrence.

Once framing effects are controlled, for example, by presenting patients with both survival and mortality information, other questions can be asked. A key one is whether the amount of information presented for choice influences patient decision making. My colleague and I have conducted two studies investigating this issue.[22] Study 1 involved having patients and physicians examine data in tables. Some patients received tables with three data points (one month, one year, and five years), others tables with six data points (one month, one year, two years, three years, four years, and five years). Of the former, 49 percent reported a preference for the long-term survival option, compared with 84 percent in those receiving more data (p = 0.00001). Physicians were provided tables with six data points and were asked to answer the same question about their own preferences for treatment. Like the patients, physicians preferred the long-term survival option (79 percent). Patients were asked which data influenced their choices. Of the patients receiving six data points, 30 percent identified the two-, three-, or four-year data. We believe this study supports the position that patients use intermediate data in their decision making. McNeil, Weichselbaum, and Pauker have previously argued that patients are typically presented two-point data: chance of surviving the surgery and chance of survival at five years.[23] Patients may need more than two data points to determine where their preferences lie.

Study 2 involved the presentation of survival data to patients *graphically*, that is, using a five-year survival curve to compare the survival and mortality rates for two treatment options. We compared a "limited explanation" involving discussion of three key data points with an "extensive explanation" involving five key data points. Of patients receiving extensive explanation, 44 percent changed their previously specified intended treatment choice, compared with 13 percent of those given limited explanation (p = 0.00006). Of the patients receiving extensive explanation, 57 percent reported either medium-term data or the average life expectancy for the five-year period covered by the curve as most influencing their decision making. Of patients receiving limited explanation, 78 percent reported only endpoint (year 0 or year 5) data as most influencing their preferences.

Thus, in both studies patients reported using medium-term data. This result becomes very problematic. The first problem arises if physicians supply only short-term (at and shortly after the time of the intervention) and long-term (five years after the intervention) information, despite the availability of medium-term data. This is not enough data for patient decision making. The second problem arises when there is no medium-term data reported in the peer-reviewed medical literature. The third problem arises when medium-term data is not even collected during a clinical trial.

Edwards et al. have described this phenomenon of patient change in choice behavior as one of increased wariness: "patients appeared to become more cautious when presented with more data."[24] In contrast, my colleagues and I believe that this change in choice behavior is due to patients' use of more data points in their decision making. Our interpretation here parallels that of McNeil et al., who argue that some patients make decisions on the basis of short-term data while others make decisions based on long-term data.[25] We extend this view by arguing that some patients base their decisions on short-term data, others on long-term data, and still others on medium-term data.

In all of the cases so far, we have considered decision making from an individual, not a societal, perspective. However, decision making for the patient becomes much more elusive when it is not clear from whose perspective scientific data in an information message are being presented. Adrian Edwards and Hilda Bastian call attention to "the ethical implications of presenting information in different ways to achieve different effects."[26] A key area for scrutiny is the approach that fails to distinguish whether the whole enterprise of informing a patient is for the *patient's good* (as valued by the individual patient) or *society's good* (as valued by someone who purports to represent that society). Presenting scientific data with *social nuance* regarding socially desirable decisions becomes an even more problematic framework for risk communication and decision making under the rubrics of *individualized decision making* and *patient autonomy.* The mix of scientific evidence regarding best outcomes with nuance or even data regarding one or more societal perspectives is also part of the new medical conversation.

Notes

1. D. J. Mazur and D. H. Hickam, "Patients' Preferences for Risk Disclosure and Role in Decision Making for Invasive Medical Procedures," *Journal of General Internal Medicine* 12, no. 2 (February 1997): 114–17.

2. D. J. Mazur and D. H. Hickam, "Patients' Interpretations of Probability Terms," *Journal of General Internal Medicine* 6, no. 3 (May/June 1991): 237–40.

3. S. Woloshin et al., "Assessing Values for Health: Numeracy Matters," *Medical Decision Making* 21, no. 5 (September/October 2001): 382–90, quoted from p. 382.

4. A. Tversky and D. Kahneman, "The Framing of Decisions and the Psychology of Choice," *Science* 211 (January 30, 1981): 453–58.

5. See G. B. Chapman and A. S. Elstein, "Cognitive Processes and Biases in Medical Decision Making," in *Decision Making in Health Care: Theory, Psychology, and Applications,* ed. G. B. Chapman and F. A. Sonnenberg (Cambridge: Cambridge University Press, 2000), 192.

6. A. Edwards et al., "Presenting Risk Information: A Review of the Effects of 'Framing' and Other Manipulations on Patient Outcomes," *Journal of Health Communication* 6, no. 1 (January/March 2001): 61–82.

7. D. J. Mazur and D. H. Hickam, "Patient Preferences for Management of Localized Prostate Cancer," *Western Journal of Medicine* 165, nos. 1–2 (July/August 1996): 26–30.

8. Edwards et al., "Presenting Risk Information," 73.

9. Edwards et al., "Presenting Risk Information," 75.

10. B. J. McNeil et al., "On the Elicitation of Preferences for Alternative Therapies," *New England Journal of Medicine* 306, no. 21 (May 27, 1982): 1259–62.

11. I. P. Levin, S. L. Schneider, and G. J. Gaeth, "All Frames Are Not Created Equal: A Typology and Critical Analysis of Framing Effects," *Organizational Behavior and Human Decision Processes* 76, no. 2 (November 1988): 149–88.

12. Levin, Schneider, and Gaeth, "All Frames Are Not Created Equal," 159–60.

13. Levin, Schneider, and Gaeth, "All Frames Are Not Created Equal," 167–68. The study referred to is B. E. Meyerowitz and S. Chaiken, "The Effect of Message Framing on Breast Self-Examination Attitudes, Intentions, and Behavior," *Journal of Personality and Social Psychology* 52, no. 3 (March 1987): 500–10.

14. Levin, Schneider, and Gaeth, "All Frames Are Not Created Equal," 174, 182.

15. J. Davis, "Riskier Than We Think? The Relationship Between Risk Statement Completeness and Perceptions of Direct to Consumer Advertised Prescription Drugs," *Journal of Health Communication* 5 (2000): 349–69, quoted from p. 353.

16. Davis, "Riskier Than We Think?" 358–59.

17. Davis, "Riskier Than We Think?" 360–62.

18. R. A. Bell, R. L. Kravitz, and M. S. Wilkes, "Direct-to-Consumer Prescription Drug Advertising and the Public," *Journal of General Internal Medicine* 14, no. 11 (November 1999): 651–57; see pp. 654–55.

19. Davis, "Riskier Than We Think?" 359.

20. On medical concerns, see J. R. Hoffman and M. S. Wilkes, "Direct-to-Consumer Advertising of Prescription Drugs: An Idea Whose Time Should Not Come," *British Medical Journal* 318 (1999): 1301–02; on ethical concerns, see J. Feiser, "Do Businesses Have Moral Obligations Beyond What the Law Requires?" *Journal of Business Ethics* 15 (1996): 457–68.

21. "Prescription Drug Advertising," 21 *Code of Federal Regulations (2000)* chap. 1 (April 1, 2001 edition), §202.1.e.5.ii.

22. D. J. Mazur and D. H. Hickam, "Treatment Preferences of Patients and Physicians: Influences of Summary Data When Framing Effects Are Controlled," *Medical Decision Making* 10, no. 1 (January/March 1990): 2–5, and "The Effect of Physician Explanations on Patients' Treatment Preferences: Five-Year Survival Data," *Journal of General Internal Medicine* 9, no. 5 (July/September 1994): 255–58.

23. B. J. McNeil, R. Weichselbaum, and S. G. Pauker, "Fallacy of the Five-Year Survival in Lung Cancer," New England Journal of Medicine 299, no. 25 (December 21, 1978): 1397–1401.

24. Edwards et al., "Presenting Risk Information," 73.

25. McNeil et al., "On the Elicitation of Preferences for Alternative Therapies."

26. A. Edwards and H. Bastian, "Risk Communication: Making Evidence Part of Patient Choice," in *Evidence-Based Patient Choice: Inevitable or Impossible?* ed. Edwards and G. Elwyn (Oxford: Oxford University Press, 2001), 153.

14

The Move toward Providing Patients with "More Information" of "Different Types"

THE MAJOR FOCUS OF THE JUDICIAL DOCTRINE of informed consent has been on risk disclosure to patients. Other authors have argued, from quite disparate perspectives, that the narrow focus on physician disclosure needs to be broadened and strengthened for issues related to patient care.

Recently authors have argued that informedness in clinical care should mirror informedness in clinical research.[1] According to the *Code of Federal Regulations*, individuals being recruited into scientific protocols must be informed of the nature of the research, the research procedure to be followed, its risks, the fact that the research may not benefit them, and the subject's right to withdraw his or her participation from research at any time.[2] In the area of risk disclosure in research consent, subjects need to be informed of the *reasonably foreseeable* risks and discomforts, and that the particular research diagnostic or therapeutic intervention may involve risks that are currently unforeseeable.

Yet major questions are not covered by current federal regulations, including defining precisely what is to be included in the concept of *reasonably foreseeable risk*. In addition, except for specifying that the concepts need to be translated into lay language, there is little attention in current federal regulations to what risk language is to be used in informed consent forms.

Recently the National Bioethics Advisory Commission (NBAC), in their report *Ethical and Policy Issues in Research Involving Human Participants*, examined the oversight system for protecting the rights and welfare of research participants. As NBAC notes, "the term *subject* is generally used in scientific disciplines in which studies involve experimenting on humans."[3] Robert Veatch likens the subject to "'research material' . . . a useful substance necessary to complete the research objectives."[4] NBAC observes that "some individuals who participate in research find the term *subject* offensive," but that "no other term more accurately portrays the relationship and the unequal balance of power between the investigator and the individual in research." While NBAC advocates the use of the term "human participants" to describe individuals agreeing to enroll in a clinical research study, three NBAC commissioners argue that

> it is premature to adopt the term, because today too many patients and volunteers who are enrolled in research studies are still not free and equal participants in the research; indeed changing the term could send a false signal that less vigilance is needed to protect human subjects or that investigators and IRBs [Institutional Review Boards] need not expend further effort to move to a system in which the people being studied are truly research participants.[5]

Here we see some of the tensions that still exist in clinical research involving human participants.

An additional problem in research on human subjects is that patients often fail to distinguish between their participation in clinical research and the clinical care they receive. This distinction between clinical care and clinical research is further complicated when the patient's clinician is also the principal investigator who is recruiting the patient into the investigator's research study. Paul Appelbaum writes that research projects

> have been criticized for failing to distinguish clearly between the benefits and risks associated with research and those that would be obtained if the potential subject sought ordinary clinical care. Risks in general sometimes seem to be downplayed. Both problems may be exacerbated when researchers are also in charge of subjects' clinical treatment.[6]

In addition, research subjects may need to be periodically reminded that they are *still enrolled in a research trial* and *not receiving clinical care*. There should be continuing efforts to reaffirm a subject's participation in a clinical research study. Dave Wendler and Jonathan Rackoff distinguish four types of continuing consent:

- reconsent
- ongoing consent
- reaffirmation of willingness to participate
- dissent.[7]

This *full range of alternatives* includes different *types* of alternatives, especially in the area of therapy. The 1982 President's Commission for the Study of Ethical Problems in Medicine and Biomedical and Behavioral Research argued that physicians should share their opinions with their patients on topics that patients were interested in—for example, alternative (nonmedical) therapies.[8] The main focus of the commission was to shift discussions away from *severe adverse outcomes at low probabilities of occurrence* (which it took to be implicit in the judicial concepts of informed consent as formulated in the United States) to other areas of interest to patients. Ashcroft, Hope, and Parker argue that patients "should be given information relevant to choosing a drug treatment, and not simply information about the drug which the doctor chooses."[9] This echoes the contention of the 1982 President's Commission that patients also want information about how to reason about a medical choice.

Once alternatives are brought more into focus in patient–physician communication, one can immediately see that what is involved is not simply information disclosure but approaches to weighing and judging that information. A simple schema for considering, weighing, and judging information about risk, benefits, and alternatives can be captured in three stages.

1. There is information disclosure by the physician.
2. There is an assessment of patient understanding of the information, the patient's level of comfort related to working with the disclosed information, and a direct answering of any questions the patient may have.
3. There are discussions between a patient and his or her physician about alternative ways of reasoning about information.

However, discussion of alternatives (in both senses—alternatives to treatment and alternative ways of reasoning about choices) as opposed to simple risk disclosure also brings into question the very information being disclosed to patients. This itself becomes a focus of debate, because it raises the issue of *evidence* (and the quality of the evidence) on which information disclosed by physicians is based.

The new framework of information now consists of the following:

- the patient's understanding of the information disclosed, in lay language;
- the information disclosed by the physician in terms of the alternatives and the evidence behind the alternatives;
- the question of the evidence behind the physician's information, and of the quality of that evidence in terms of its accuracy;
- the patient's understanding of the evidence behind the information, and its impact (if any) on the patient's decision;
- the issue of the basis on which the patients should weigh the alternatives.

As we begin to see, the notion of information viewed within this broader framework demands a conceptual shift in the way both patients and physicians view their relationship. It demands that more time be available for discussion. It involves a conceptual shift—especially on the part of physicians—to begin discussing new concepts, including the evaluation of scientific evidence with which physicians themselves may be unfamiliar. In fact, some physicians may actively disagree with the emphasis of *scientific evidence* over *clinical experience* as a basis for forming the foundation of information in the patient–physician relationship.

But from our perspective, the above framework shows the complexity of issues involved in the communicating and understanding of information by patients, physicians, and others, particularly those interested in developing the short information message for patients and consumers. What is needed is a means to establish the *priority* of information for patient evaluation, from least important to most important. Is it reasonable to put patients in the position of deciding which types of information are most important for their individual decision making? Ashcroft, Hope, and Parker argue that "patient choice will not be best served by simply giving the patient 'the evidence' and allowing him or her to get on with making his or her 'individual choice.'"[10] Yet there are many questions regarding the source(s) of this new evidence.

However, there are other issues involved in clinical care beyond the discussion of available alternatives—most notably, prognostic information as it relates to the patient with a medical condition or disease. We have so far been discussing risk and benefit disclosures, but we have said little regarding the data on which this information for disclosure is based. Is the information in the published medical literature more important to the patient? Or is the physician's personal experience with the screening, diagnosis, and management of the medical condition or disease?

Let us take the example of the information needs of cancer patients. Mesters et al. expect the cancer patient's need for information to be high in three phases of treatment:

- the period immediately before the medical treatment (discovery and diagnostic phase);
- the treatment (at the hospital or during outpatient treatment);
- the period after treatment (returning home, recovery, residual, terminal phase).

According to the authors, "research found that the vast majority of cancer patients wanted to be given all available information about their condition, whether good or bad," and the "literature on information needs among cancer patients suggests that needs for information remain unmet for a major [portion] of the patients . . . ." The authors further note the "existence of misconceptions among health care providers regarding patients' informational needs for information." Finally, these "unmet informational needs are believed to increase emotional distress for patients (e.g., increased anxiety and depression) and subsequently to hamper patients' adjustments to their illness."[11]

Elizabeth Lamont and Nicholas Christakis, in a study of cancer patients in five hospices in Chicago, asked physicians to formulate survival estimates for their patients and to provide estimates of how confident they were in the prognostic estimates they formulated. The authors also asked whether the physicians would provide these prognostic estimates to their patients, all of whom were "near the end of life." The physicians in Lamont and Christakis's study were able to formulate prognoses for 300 of 311 patients who were available for evaluation. The authors found the following results:

- Only 37 percent of the time did physicians favor providing frank survival estimates to patients with terminal cancer referred for hospice for palliative care.
- Physicians favored providing apparently knowingly inaccurate survival estimates for 40.3 percent of patients.
- Physicians favored providing no survival estimates for 22.7 percent of patients.
- Physicians with more practice experience were nearly three times as likely to favor no disclosure over frank disclosure.[12]

Our discussions of risk, benefit, and prognostic disclosures raise other broad questions. What is the *source of the information* to be provided by physicians and other health care providers to patients? And what is the *reliability* of that source? The potential sources of information are several: published medical literature, hospitals in the region, hospitals in the patient's own community, the hospital where the patient is being treated, the group of physicians who care for patients with similar medical conditions and disease, and the individual physician or provider. Lamont and Christakis did not ask how the physicians arrived at their prognostic estimates. They could have been using one or more sources of information. For example, if physicians were aware of numbers in the published medical literature, they might then attempt to apply those numbers to the individual patient depending on his or her level of vigor, concomitant medical conditions, and so forth.

Lamont and Christakis note that the science of prognostication itself needs much more study, development, and refinement, particularly in the area of minimizing errors. As the authors observe: "What good does it do to encourage physicians to communicate information that is, after all, inaccurate?" The authors point to approaches involving "statistical algorithms, averaging prognoses made by several physicians, and consulting more experienced colleagues or textbooks." But these approaches are in their infancy. In addition, even if such a prognostic science were at hand, the issue of when in the course of treatment to give the physician's prognosis still needs to be systematically addressed.

In our research to date, we have used examples that rely on common decisions in clinical care related to chances of survival in the short, medium, and long terms. However, it is important to recognize that the choices patients face in their lives include more complex decisions, requiring even more risk information.

Fortin et al. studied forty peri- and postmenopausal women. These women reported that they had specific preferences for the format in which risk data were displayed.[13] The authors studied the responses of women to risks for coronary heart disease, hip fracture, and breast cancer, in the context of hormone replacement therapy. The authors provided patients with various graphic formats: bar graphs, line graphs, thermometer graphs, 100 representative faces, and survival curves. The authors selected formats that lent themselves to depicting risks over time.

Fortin et al. found that bar graphs were preferred by 83 percent over line graphs, thermometer graphs, 100 representative faces, and survival curves. In addition, lifetime risk estimates were preferred over ten- or twenty-year time horizons. Finally, the authors found that *absolute risks* were preferred over *relative risks* and *numbers needed to treat*.[14] The authors did not conduct any measurements related to their patients' understanding of data in any of the preceding formats, to whether the patients were interpreting the data correctly. This was a study of patient preferences for *formats* of scientific information only.

The key point regarding short information messages is that there may not be room to display information in more than one format. Yet researchers have suggested that the display of information in more than one format is key to control framing effects.[15] There are key questions related to how much information patients want, and at what level of accuracy. Given the different approaches physicians can use to generate prognostic estimates, to what extent do patients want to see and understand the evidence behind the numbers?

It is also important to determine the challenges patients face in understanding numbers. We have discussed framing effects as an introduction to problems that can be encountered with short information messages. Still, numbers and patients' reported preferences regarding numerical information are also active areas of research. In my research, those patients who preferred physicians to use verbal chance terms in explaining risks reported different reasons for their preference: "I don't understand numbers"; "I don't trust numbers"; "numbers can be manipulated." Recent research has also shown that many patients are unable to perform simple mathematical operations using numbers.[16] So patients' distrust of numbers may come from their distrust of their own abilities to understand and work with numbers or from skepticism about the accuracy of the numbers.

Robert Makus's notion that medical information on the Internet is egalitarian assumes that two criteria are satisfied.[17] First, it assumes that individuals have the tools and skills to access the information as well as the ability to understand and work with it. Second, it assumes that the information on the World Wide Web is translated into nonprofessional language that is understood by patients. Yet no matter how much information one provides, there are issues concerning the time available to patients and physicians in short clinic sessions to talk about information in a fashion that is satisfactory to both parties.

Finally, short advertisement messages become an issue for the physician as the learned intermediary who is ultimately responsible for writing a prescription for the medication being advertised. This responsibility extends directly to the physician's obligations under the respective judicial doctrine of informed consent in the state in which the physician practices, that is, the judicial professional standard and the judicial reasonable person standard regarding informed consent for prescription drugs.

The Agency for Healthcare Research and Quality and the Kanter Foundation argue that the way to improve health care decision making is by "giving people information—including print and Internet resources—based on scientific research."[18] Their focus is on information that is "science-based and reputable"; individuals should "look at the latest and best studies" and "exercise caution," particularly with respect to

- phrases such as "scientific breakthrough," "miraculous cure," "exclusive product," "secret formula," and "ancient ingredient";
- the use of "medical-ese"—impressive technical terms to disguise a lack of good science;
- case histories from "cured" consumers claiming amazing results;
- a laundry list of symptoms the product cures or treats;
- the latest trendy ingredient making headlines;
- a claim that the product is available from only one source, for a limited time;
- testimonials from "famous" medical experts;
- a claim that the government, the medical profession, or research scientists have conspired to suppress the product.

This kind of language can be used in short information messages, and both the Agency for Healthcare Research and the Federal Trade Commission urge individuals to beware of it.

Note

1. See I. Chalmers and R. Lindley, "Double Standards in Informed Consent to Treatment," in *Informed Consent: Respecting Patients' Rights in Research Teaching and Practice*, ed. L. Doyal and J. S. Tobias (London: British Medical Journal Books, 2000).

2. 45 *Code of Federal Regulations* Subtitle A (October 1, 2001 edition), §46.116.

3. National Bioethics Advisory Commission (NBAC), *Ethical and Policy Issues in Research Involving Human Participants*, vol. 1 (Bethesda, Md.: The Commission, August 2001), 33.

4. R. M. Veatch, *The Patient as Partner: A Theory of Human-Experimentation Ethics* (Bloomington: Indiana University Press, 1987), 4.

5. NBAC, *Ethical and Policy Issues*, 1: 32–33.

6. P. S. Appelbaum, "Rethinking the Conduct of Psychiatric Research," *Archives of General Psychiatry* 54, no. 2 (February 1997): 117–20, quoted from p. 117.

7. D. Wendler and J. Rackoff, "Consent for Continuing Research Participation: What Is It and When Should It Be Obtained?" *IRB: Ethics & Human Research* 24, no.3 (May/June 2002): 1–6.

8. U.S. President's Commission for the Study of Ethical Problems in Medicine and Biomedical and Behavioral Research, "Informed Consent as Active, Shared Decision-making," in *Making Health Care Decisions: A Report on the Ethical and Legal Implications of Informed Consent in the Patient–Practitioner Relationship*, vol. 1 (Washington, D.C.: President's Commission for the Study of Ethical Problems in Medicine and Biomedical and Behavioral Research, 1982), 15–39.

9. R. Ashcroft, T. Hope, and M. Parker, "Ethical Issues and Evidence-Based Patient Choice," in *Evidence-Based Patient Choice: Inevitable or Impossible?* ed. A. Edwards and G. Elwyn (Oxford: Oxford University Press, 2001).

10. Ashcroft, Hope, and Parker, "Ethical Issues," 63.

11. I. Mesters et al., "Measuring Information Needs Among Cancer Patients," *Patient Education and Counseling* 43, no. 3 (June 2001): 253–62, quoted from p. 253.

12. E. B. Lamont and N. A. Christakis, "Prognostic Disclosure to Patients with Cancer Near the End of Life," *Annals of Internal Medicine* 134, no. 12 (June 19, 2001): 1096–1105.

13. J. M. Fortin et al., "Identifying Patient Preferences for Communicating Risk Estimates: A Descriptive Pilot Study," *BMC Medical Informatics and Decision Making* 1 (2001), at www.biomedcentral.com/1472-6947/1/2 (accessed July 7, 2002).

14. For a fuller discussion of the concept of *numbers needed to treat*, see S. Ebrahim, "The Use of Numbers Needed to Treat Derived from Systematic Reviews and Meta-Analysis: Caveats and Pitfalls," *Evaluation and the Health Professions* 24, no. 2 (June 2001): 152–64; L. Smeeth, A. Haines, and S. Ebrahim, "Numbers Needed to Treat De-

rived from Meta-Analyses: Sometimes Informative, Usually Misleading," *British Medical Journal* 318 (June 5, 1999): 1548–51; and P. J. Wiffen and R. A. Moore, "Demonstrating Effectiveness: The Concept of Numbers-Needed-to-Treat," *Journal of Clinical Pharmacology* 21, no. 1 (February 1996): 23–27.

15. See R. Ashcroft, T. Hope, and M. Parker, "Ethical Issues and Evidence-Based Patient Choice," in *Evidence-Based Patient Choice: Inevitable or Impossible?* ed. A. Edwards and G. Elwyn (Oxford: Oxford University Press, 2001), 61–62.

16. I. M. Lipkus, G. Samsa, and B. K. Rimer, "General Performance on a Numeracy Scale among Highly Educated Samples," *Medical Decision Making* 21, no. 1 (January/February 2001): 37–44.

17. R. Makus, "Ethics and Internet Healthcare: An Ontological Reflection," *Cambridge Quarterly of Healthcare Ethics* 10, no. 2 (Spring 2001): 127–36.

18. Agency for Healthcare Research and Quality and the Kanter Family Foundation, "Now You Have a Diagnosis: What's Next? Using Health Care Information to Help Make Treatment Decisions" ([Rockville, Md.]: Agency for Healthcare Research and Quality, 2000).

IV

Communicating Risk-Benefit Information in the Future

15

Egalitarian Approaches to Information

ROBERT MAKUS OBSERVES that while a physician may tell a patient that only three treatments are available for a particular condition, the individual could go to the World Wide Web, where some twenty other treatments (some having undergone rigorous scientific study, others not) might be available, treatments the physician never mentioned. Makus argues that one of the clear advantages of the development of what he describes as "Internet medicine" and websites devoted to health issues is that "information previously parceled out by one's doctor is now easily available to anyone with access to a computer." By judicious use of such websites, an individual can find out about the latest, most sophisticated research on medical conditions and diseases, as well as about alternatives to traditional methods of treatment. Makus argues that this "brings a new egalitarianism to healthcare . . . because the opportunity to find out about medical issues is no longer determined by the patient's final resources or proximity to [major medical centers]."[1]

Of note is the issue of *egalitarian* approaches in relation to *full disclosure.* From our standpoint, full disclosure would entail not only descriptive information about such research on medical conditions, but also the *scientific status* and *quality of scientific evidence available* related to that information. It is not clear whether Makus's egalitarian approach would include information about treatment recommendations that may not have been fully explored, and might not ever be fully explored, depending on the scientific merit of the medical intervention and its scientific status at a particular time.

As the Agency for Healthcare Research and Quality notes, there is an assumption that the websites being accessed should represent reputable opinions.[2]

But who is to judge the strength of this opinion, and what are the criteria that should be used to judge? Makus's reference to the opinion of major medical centers is particularly important here because these centers have their own websites that provide certain levels of information. In addition, there are websites developed by health and medical professional organizations.

The issue of who is in possession of the knowledge to interpret the scientific information released on the World Wide Web, and why it is being released now, can be broken down into a set of nine questions.

- Who developed the data?
- Were the data developed by experienced researchers?
- Was the study designed appropriately?
- Who is releasing the data?
- Why is the World Wide Web the chosen forum for release of the data?
- Are the data published in a peer-reviewed medical journal?
- Who has interpreted the data?
- Who has reviewed the data in terms of the quality of the interpretation?
- Are there any conflicts of interest that might impact the way the study was designed and the data interpreted?

This approach to understanding scientific data might prove useful in terms of a strategy for consumers and patients retrieving medical information on the World Wide Web.

Two questions need to be asked of Makus's view of egalitarian information.

1. Is the data on the World Wide Web derived from an appropriately designed scientific study?
2. If the answer is yes, is the data appearing on the World Wide Web accurate in terms of its interpretation?

Answers to both questions fall into what might be termed the *technical domain*. In a sense, no matter where information appears—in a short information message or in a website—it has largely been kept in the domain of the individual with enough technical skills and experience to interpret the available scientific data. These individuals may include physicians, scientists, epidemiologists, statisticians, economists, and decision analysts among others. A less than full disclosure of information does not necessarily mean that information has been intentionally kept away from patients. Rather, the information may have been derived from inadequately designed research studies. We have already seen this point in our discussion of Elizabeth Lamont and Nicholas Christakis's study of physician prognostic estimates in chapter 14. Lamont and

Christakis found that (1) physicians with more practice experience favored no disclosure over frank disclosure, and (2) physicians with previous experience in the palliative care of dying patients favored disclosure of knowingly inaccurate pessimistic prognoses rather than frank prognoses. The question to be asked related to both findings is, why are physicians behaving in this manner?

What is needed is a new research study to try to answer this question. If Lamont and Christakis's study were placed on a website, a consumer might mistakenly assume that *speculations* regarding the data are *findings of the study*. A follow-up study should be formulated to attempt to answer the question of why physicians behave as they indicated they would in Lamont and Christakis's study.

The technical domain requires a firm understanding of study design and a high level of skill in interpreting the quality of the available data and recognizing when there is not enough scientific data available to make a decision. Needed skills include an understanding of epidemiology, statistics, and other technical areas. There may be arguments about who actually has the needed skill to make interpretations in the technical domain. But no matter how *technical* the interpretation of data becomes, it must at some level be communicated to the nonscientist. Thus, the technical domain demands at least four types of skill:

- skill at designing scientific studies;
- skill at interpretating the data resulting from high-quality scientific studies;
- skill at communicating between the scientist, the physician, and the technician;
- skill at translating from the technical domain of the scientist to the domain of the consumer and patient.

The most difficult skill to master is that of translating from the technical domain of the scientist to the domain of the consumer and patient. The physician, by his or her direct contact with the patient in medical health care matters, is often placed in the position of learned intermediary regarding this information. The challenge of scientific information on websites is whether there will be a learned intermediary that the patient may access related to the interpretation of the technical information.

Yet it has been argued that physicians also have rights of autonomy in decision making,[3] and limits on what they can be forced or pressured to do within their role.

The question of patients using physicians as learned intermediaries related to scientific information garnered from sources like the World Wide Web must be understood in light of the pressures physicians face in their practice. Although we have focused on consumers and patients as the *targets* of the short information message and behavioral change, forces are likely operating to change physician behavior unrelated to the simple provision of information to physicians. Indeed, many approaches in the published medical literature do not advocate provision of scientific evidence to physicians as a mode of behavior change. Rather, they are based on expert models, for example, relying on the opinion of local expert physicians to influence behavior change in their colleagues. But the simple approach of providing physicians with clear scientific information from the peer-reviewed medical literature has not been successful in changing physician behavior. In addition, economic incentives have been used to change physician behavior. Such inducements (e.g., to select more expensive vs. less expensive drugs) are provided by the physician's employer and are often incorporated into the contracts that physicians sign.

While Makus argues for the egalitarian approach to information provided by the Internet, Katz claims that physicians have been too constricted in their communication of information to patients, and that the current state of physician–patient decision making is dominated by physicians' judgments as to what is best. He notes that

> formidable problems exist which require study and resolution before informed consent can ever safeguard patient autonomy and self-determination. The likely outcome of such inquiries will be to return ownership of bodies to patients and not allow caring custody to mislead physicians and patients into believing that ownership must temporarily be transferred to doctors' "discretion."[4]

I do not necessarily agree with Katz's interpretations here. I believe, for many decisions, the physician *is selected by others* to be the intermediary regarding scientific information. And, in fact, today it could be argued that the physician is becoming or already is the *economic intermediary* in terms of getting access for his or her patients to certain expensive diagnostic and therapeutic procedures. For example, Peter Ubel actually discusses physicians' "pricing of life."[5] It is not clear how many physicians are actually comfortable in their roles of *medical product intermediary* and *economic intermediary*. Future research will need to examine the training, beyond medical training, that is needed for such roles.

Notes

1. Makus, "Ethics and Internet Healthcare: An Ontological Reflection," *Cambridge Quarterly of Healthcare Ethics* 10, no. 2 (Spring 2001): 127–36, quoted from p. 127.

2. Agency for Healthcare Research and Quality, at www.ahcpr.gov (accessed July 6, 2002).

3. See R. M. Veatch, *The Patient as Partner: A Theory of Human-Experimental Ethics* (Bloomington: Indiana University Press, 1987), and *The Patient–Physician Relationship: The Patient as Partner, Part 2* (Bloomington: Indiana University Press, 1991).

4. J. Katz, "Informed Consent: Must It Remain a Fairy Tale?" *Journal of Contemporary Health Law and Policy* 10 (Spring 1994): 69–91, quoted from p. 71.

5. P. A. Ubel, *Pricing Life: Why It's Time for Health Care Rationing* (Cambridge: MIT Press, 1999).

16

Research on Communication in the Patient–Physician Relationship

P ATIENT AGENDAS IN CLINICAL CARE have been documented in terms of the issues patients bring up with their physicians regarding direct-to-consumer advertising for over-the-counter and new prescription drugs.[1] At the same time, recent research on practice guideline implementation sheds light both on aspects of patient–physician communication and on patient agendas in clinical care unrelated to drug treatment.

Let us take an example involving the use of *practice guidelines* for low back pain. In the Netherlands, Schers et al. conducted forty interviews with twenty patients who consulted for low back pain and with the twenty general practitioners who saw these patients in clinic.[2] The physicians used a guideline for how to handle low back pain in patients. Patients reported that they wanted to hear a diagnosis or expected to receive simple advice. The general practitioners said they were well informed about the guideline and mostly agreed with its content, yet they did not strictly adhere to it. Reasons for nonadherence were mainly related to patients' experiences in the past and general practitioners' interpretations of patient preferences. General practitioners stated that they were inclined to give in to patients' demands (for example, for a radiograph or a referral to a physical therapist) rather than follow the guideline. The researchers argue that implementation strategies related to guidelines "should aim at training physicians in communication skills, especially about subjects for debate, where patients' beliefs and experiences color their expectations."[3]

Sox, Margulies, and Sox tested the hypothesis that routine diagnostic studies and laboratory tests affect patients even if those tests have no diagnostic value.[4] The authors measured clinical outcomes of 176 patients thought clinically to have nonspecific chest pain. Patients were randomly allocated either to have a routine electrocardiogram and serum creatine phosphokinase test or to have all diagnostic tests withheld. Fewer patients in the test group (20 percent) reported short-term disability after the visit than patients in the no-test group (46 percent; p = 0.001). Logistic discriminant analysis confirmed that the use of diagnostic tests was an independent predictor of recovery. Patients in the test group felt that care was "better than usual" more often (57 percent) than patients in the no-test group (31 percent; p = 0.001). After the index visit, the two groups were equally worried about serious disease and equally sparing in their use of other medial care for chest pain.

What conclusions can be drawn from studies like these? First, it appears that simply informing patients about back pain and chest pain may not have an immediate influence on patients' agendas. Second, as Schers et al. recommend, physicians should be trained in communication to better engage the patient in debate. Third, Sox, Margulies, and Sox demonstrate that patients rate their clinical care as better if they are provided with confirmatory diagnostic studies and laboratory test results in addition to physician counsel. Fourth, Schers et al. argue beyond diagnostic testing to the provision of a consultation to obtain a therapeutic modality, physical therapy with hands-on care.

Notes

1. For example, see F. Sheftell, D. Dodick, and R. Cady, "Direct-to-Consumer Advertising of OTC Agents under Current FTC Regulations: Concerns and Comment," *Headache* 41, no. 6 (June 4, 2001): 534–36.

2. H. Schers et al., "Implementation Barriers for General Practice Guidelines on Low Back Pain: A Qualitative Study," *Spine* 26, no. 15 (August 1, 2001), at www.spinejournal.com (accessed August 8, 2002).

3. Schers et al., "Implementation Barriers."

4. H. C. Sox Jr., I. Margulies, and C. H. Sox, "Psychologically Mediated Effects of Diagnostic Tests," *Annals of Internal Medicine* 95, no. 6 (December 1981): 680–85.

17

The Longer Information Message: Toward a Fuller Understanding of the "Range of Information" Being Discussed

WE HAVE SO FAR LIMITED OUR DISCUSSION to the *shorter information message* (e.g., in contrast to *full disclosures*), yet there is a broad range of information that patients may want and need related to the diagnosis and treatment of a condition, only some of which deals with risk, benefit, and prognosis.

The Toronto Informational Needs Questionnaire on Breast Cancer (TINQ-BC) consists of fifty-one items measuring five subscales on information needs in the areas of

- disease;
- investigative tests;
- treatment;
- physical state;
- psychosocial state.[1]

Degner et al. studied patients' preferences for information by presenting subsets of two information needs items from a total of nine, including

- chances of cure;
- communicability of disease;
- treatment options;
- family risk;
- adverse effects;
- home self-care;
- impact on family;

- impact on social activity;
- impact on sexuality.[2]

The Patient Learning Needs Scale (PLNS) measures discharge information needs on a forty-item self-report scale with five subscales:

- concern, support, and care in the community;
- medication;
- treatment and activity of daily life;
- complications and symptoms;
- illness-related concerns.[3]

Finally, the Patient Information Need Questionnaire (PINQ) consists of items on the following topics:

- the patient's present condition;
- the cause of the patient's illness;
- the possible course of the patient's illness;
- survival rates for the patient's illness;
- possible consequences of the patient's illness;
- the benefits or goals of the treatment;
- possible side effects of the treatment, such as tiredness or an increased burden of care falling on someone else;
- treatment procedures;
- what the patient can do (or is allowed to do) in his or her situation (e.g., work, hobbies, food and drink);
- how the patient is to find his or her way around the hospital;
- the best way to talk or interact with the patient's physician;
- opportunities to get immediate help if the patient experiences problems;
- where the patient can get good educational materials or literature about his or her illness or treatment;
- prostheses (e.g., wigs);
- how to keep or become physically fit (exercises and diet).[4]

These related, yet different, approaches to the categorization of information begin to demonstrate the types of information that patients may feel they need at some or all of the phases of their medical condition or disease.

Information should be evaluated against the background knowledge of the recipient. Currently, few studies assess such background knowledge. In addition, it is important to recognize that the sick state itself may influence memory, cognition, and even recognition.[5]

Ziegler et al. found that desire for maximum information about adverse effects of medications was significantly correlated with previous frequent experience of adverse effects of medications (p < .001).[6] Thus, a patient's past experience with an adverse outcome may be used to predict that the patient wants maximum information from physicians. The researchers also found that 73.4 percent of individuals believe a physician is never justified in withholding information about adverse outcomes.

The real issues here can be captured in a set of questions. First, does the patient want the information? Second, what are the categories of information that patients want to know? Third, how much of the information in each category does the patient want?

We have reviewed cases in which researchers have studied what information patients want regarding the chance of a side effect, complication, or adverse outcome occurring at a particular level of severity. But rarely is it asked *who* the patient feels is the appropriate individual or group to provide the needed information.

Computer information and/or decision support systems may be able to take up much of the burden for patients who are familiar with the Internet. In my own practice, when I begin to discuss the adverse outcomes of a new medication I am recommending, some of my patients stop me cold and say that they'll access that information on the Web when they get home. While more and more information is being added to the World Wide Web on an hourly basis, what has been lacking is the development and deployment of *systems to monitor the quality of that information* in terms of its scientific accuracy and accuracy of interpretation. Also lacking are systems to monitor the way the scientific information is being expressed and translated from the technical language of the scientist to the non-technical language of the non-scientist.

It is often assumed that there is a much clearer distinction between what a "professional" wants and what a patient wants in terms of information. For example, Ziegler et al. note that "patients expressed the desire for all information about risks for adverse effects, no matter how rare, which seems unrealistic to professional personnel."[7] Yet physicians have more background knowledge and more experience with the occurrence of side effects in others, and this may influence what information a physician would like to have himself or herself for decision making. In addition, little research has been done on what information physicians actually do want.

There are also issues of individual preference in relation to background knowledge and personal experience. For example, some patients may view an extensive listing of side effects, complications, and adverse outcomes as something they want to work through when they are about to take a new drug, whereas other patients may have much fewer requirements for information.

As we enter the arena of studies in risk communication, it is important to recognize the constraints of time on the patient–physician relationship. Studies in risk communication, for the most part, are done in the outpatient setting. Thus, the fragments of conversations that risk communication researchers access are from very time-constrained patient–physician sessions. One of the biggest problems in the area of shared decision making is the constraint on alternatives presented to patients. The problem is that while some constraints are real and widely shared, others are a product of how a particular medical institution is attempting to control its expenditures on patient care.

The courts have struggled with the notion of a full disclosure of risks because of the difficulty in determining when full disclosure has been made. Some states have developed teams of patients and attorneys who sit in judgment of the adequacy of information contained in informed consent forms.[8] But it is not clear in any scientific sense how well these teams represent patients in clinics, hospitals, and intensive care units.

Today, because of economic constraints, physicians must struggle with the notion of full disclosure of alternatives. It becomes particularly difficult for physicians to give educated opinions in the area of nonmedical interventions— for example, complementary alternative medical care (CAM therapies)—where there may be little scientific evidence available and little regulation of products or providers.[9]

Mark Tonelli argues that knowledge from five areas needs to be incorporated into each medical decision: empirical evidence, experimental evidence, physiologic principles, patient and professional values, and system features.[10] Tonelli notes that proponents of evidence-based medicine have clearly acknowledged one aspect of the gap between *empirical evidence* and *clinical practice*: "the part that requires the consideration of values, both patient and professional, prior to arriving at medical decisions."

> Not as clearly recognized, however, is the gap that exists [because] empirical evidence is not directly applicable to individual patients, as the knowledge gained from clinical research does not directly answer the primary clinical question of what is best for the patient at hand. . . . Proponents of evidence-based medicine have made a conceptual error by grouping knowledge derived from clinical experience and physiologic rationale under the heading of "evidence" and then have compounded the error by developing hierarchies of "evidence" that relegate these forms of medical knowledge to the lowest rungs. . . . [T]hese latter forms of medical knowledge differ in kind, not degree, from empirical evidence and do not belong on a graded hierarchy. As they differ in kind, these other forms of medical knowledge can be viewed as complementary to empirical evidence and their incorporation is necessary to overcome the intrinsic gap [between scientific evidence and clinical practice].[11]

What is still lacking, however, is the scientific push needed to begin to attempt to interrelate these different forms of medical knowledge in a coherent scientific framework.

Medical/Scientific vs. Socioemotional Issues in the Clinic Visit

The clinic visit can be seen from both a medical/scientific dynamic of conversation between a patient and a physician and from a socioemotional dynamic. These give different views of what goes on in a clinic visit. For example, the court-based view of informed consent argues that where informed consent is needed prior to a medical intervention, the medical conversation is about a description of the nature of the procedure, its risks, and the alternatives to the procedure. A key distinction separates the clinical conversation into two segments. First, there is the area of medical/scientific explanation. Second, there is the area of socioemotional inquiry and rapport building.

Until now, research involving patient and physician interactions in clinical settings has focused more on inquiries related to socioemotional issues than on medical/scientific explanation. We must likewise consider the institutional time limits that restrict conversations in the patient–physician relationship. In the United States, clinic visits last sixteen to seventeen minutes, with little difference in time between family practice, general practice, internal medicine, or obstetrics/gynecology visits.[12] With this understanding, we will now approach the type of research that has been conducted in such settings, where time is defined, for the most part, by the institution in which the physician practices.

Researchers in communication use the following types of categories. Some researchers group categories into two main clusters, affective behavior and instrumental behavior.[13] The term *instrumental behavior* is used to capture more technically based skills used to address patients' problems and concerns. *Instrumental behavior* is divided into categories and includes information about clinical disease and its symptoms and signs, diagnostic testing, and therapies or management strategies. The terms *socioemotional* and *affective behavior* are used to capture issues related to physicians' attempts to create a strong interpersonal relationship and rapport with patients.

But there are possible shortcomings related to such categorization frameworks. The very choice of categorization schemas influences what is being sought in research. For example, the *instrumental* category needs to be expanded to include additional issues of false positivity and false negativity of testing and evidence as a topic for discussion in the patient–physician relationship. Categorization frameworks can manipulate what is going on in research related to patient–physician discussion by looking for changes in a certain set of concepts and behaviors.

The research methodology of categorizing text strings involves first obtaining informed consent from both patients and physicians to audiotape medical conversations, then transcribing those tapes and analyzing strings of text (text strings), and finally placing those strings in various categories. Researchers studying patient–physician conversations face a number of important problems with this methodology. First, there are problems with categorizing text strings in terms of their meaning in the patient–physician conversation. If the conversation takes place within a short time interval, the text strings will also be short and may become difficult to classify within a relevant category. There may be differences of opinion on whether a particular text string should be counted as an item in a particular category, or whether the string is simply ambiguous and incapable of being categorized. Likewise, it may be difficult to separate categories and subcategories within the communication researchers' framework, for example, to decide what is a socioemotional vs. a psychological subcategory. This sorting influences any interpretation of what is actually going on in the patient–physician relationship. Second, there are problems with obtaining the appropriate informed consent regarding the audiotaping or audiovisual taping of a patient–physician conversation. The very act of taping an encounter may alter what is said, or how it is said. Finally, there is the problem of interpreting what these communication behaviors mean, because the text strings are analyzed without any structured interviews asking the patient or physician what they meant by the words.

These problems exist in the nontechnical and non–scientifically based medical conversation. But these problems will expand exponentially when the text strings are made of more scientifically and technically based terms and concepts. Researchers will need to expand their categorization frameworks to capture these new scientific and technical issues as we enter the new medical conversation.

Given institutional time constraints on the patient–physician conversation, it is important to examine the use of the time that is available. This is conditioned by at least three factors: (1) the patient's agenda, (2) the physician's agenda, and (3) the medical organization or institution's agendas. At times, there may be an overlap of agendas.

Let us look at the example of a woman fifty years of age who may be concerned about the media's attention to the fact that there are no clear consensus (at the time of this writing) on whether women between the ages of forty and forty-nine should be screened. The patient's physician may have perspectives to share, and the medical institution may have a mandate that breast cancer screening must be done on adult women of specific ages, not including the

40–49 age range. Here, each participant in the debate—patient, physician, and medical institution—has the same topic of concern for the clinic visit, but they may have widely disparate views of how the decision making should go. At other times there may be more disparate agendas at work. Within this array of agendas, there are questions regarding how to approach decision making between patients and physicians. For our purposes, *decision making* relates to the weighing and balancing of information related to alternatives and to choosing a value measure or set of value measures to use in doing so. Decision making refers to the consideration, ranking (where possible), and prioritization of risk and benefit according to specifiable values. It relies on the judgment of individuals regarding these states as a whole or regarding aspects of qualities of these states.

A problem that comes up is that, many times, rankings (prioritizations) depend on *state descriptions,* and on how these states are conceived and described. On the patient's side, there is always the potential problem of misinterpretation and failure to ask questions. On the physician's side, there may also be a lack of patient support, a lack of consideration of the patient's perspective, and a failure to appreciate that a patient may have trouble formulating questions. There need to be ways to help patients understand what is going on in the patient–physician conversation.

So decision making is not only built on optimal information but also on a type of *global understanding* of what is going on, that is, an understanding of who will be doing what, the competencies of all involved, and where and when things will occur. There is also the *specific understanding* of the states and qualities involved in weight and balancing, judging, ranking, and prioritizing. Much less attention has been placed on global understanding than on specific understanding. Because of the time constraints within the patient–physician conversation, there is much less research directed at the notion of what it means to be misinformed, the problems of interpretation and misinterpretation, and the ways to identify, correct, or prevent the manipulation of information by any party.

I recommend that we place more emphasis on the medical/scientific, or instrumental explanatory behaviors that are present or absent in patient–physician communications. These types of explanations involve issues of diagnosis, management, and treatment of disease, but they also include error rates and problems with the interpretation of testing of such diagnosis, management, and treatment frameworks within medical science. In addition, they include explanations based on the scientific evidence that forms the basis of medical judgment and the range of reasonable interpretations that underlie scientific studies, which in turn form a basis for the individual physician's judgment in a particular case.

The judicial approach may be viewed as deficient in terms of the clarity courts have provided regarding the level of detail that must be provided to patients. In addition, while the courts may want all informed consent forms to be written in language and terms the patient can understand, they have shed little light on how this translation process should take place. Any area of complex and newly developing research methodologies, where there may be a potential for more severe adverse outcomes than are recognized at present, can serve to illustrate these problems of interpretation and translation.

A case in point is the developing area of genetic testing. At issue in genetic testing is not only the individual's decision to allow tissue to be stored for future research but also the concerns of that individual's descendents and future generations. These issues are particularly problematic because of the chance that confidential information will be released. Employers and insurers could then access that information and use it in ways that negatively affect present and future generations. Again, problems can come up when there is a lack of agreement among family members on whether genetic material should be donated.

Evidence-based medicine has taken on the task of medical/scientific explanation in the patient–physician relationship with a strong preference for the disclosure and understanding of the nature of scientific evidence, its problems, and its uncertainties. The problem with descriptions is that they can be manipulated. For example, a patient could be given the following description of chest compression in cardiopulmonary resuscitation (CPR): "You don't want us to break the bones of your chest, do you?" Or a patient could receive a description of chest compression as follows: "We will have to apply pressure to your chest to help your heart pump blood and continue your blood circulation while we try to reestablish the normal beating of your heart. Would you like us to support you in this manner if you need this procedure to stay alive?" All medical/scientific explanations or instrumental explanatory behaviors can be altered and, to some extent, manipulated—intentionally or unintentionally—by the physician.

Another area of inquiry and research, that of *shared decision making*, has taken on the dual task of medical/scientific explanation and socioemotional inquiry. As we shall see, the most extensive proposals for information in the patient–physician relationship have come from proponents of shared decision making.

Let us examine shared decision making first from the perspective of a rational decision maker. I have previously argued that a framework for shared decision making includes the following nine phases.

1. giving information (disclosure);
2. gathering information;
3. understanding information;
4. understanding decision strategies;

5. choosing a decision strategy;
6. using a decision strategy;
7. tentatively deciding;
8. negotiating;
   • negotiating related to individual values;
   • negotiating related to economic constraint: the individual vs. society;
9. deciding.[14]

Here as always information has an important role in the decision process. But shared decision making has multiple components, going well beyond the judicial approach with its emphasis on disclosures about risk, benefits, and alternatives; well beyond the ethical approach of ranking harms; and well beyond the decision-analytic approach of assessing a patient's preferences for outcomes among alternatives. These may be termed *systematic* approaches to decision strategies. Still others argue for the existence and use of *intuitive* approaches to decision making. The nine-step framework I use attempts to expose patients to alternative decision strategies to help them weigh, judge, and prioritize information in their decision making. Here, the disclosure of information is not viewed as an end in itself, but rather information provision is a first step in a rational decision-making process.

Negotiating about the Decision

The term *negotiation* itself suggests that a problem has arisen in the shared decision process. Problems requiring such negotiation are of at least two types. First, there are problems related to a clash between individuals and their respective *values*. Second, there are problems related to a clash between individual values and what has been termed the *societal perspective*. One particular type of societal perspective is captured in the notion of economic constraint and the rationing of health care.

The Clash between Individual Values

Negotiation as a part of shared decision making comes up each time there is a clash between individuals based on the values they hold. For example, a patient may want a referral to a physical therapist for continued therapy for low back discomfort, while the physician may recommend a set of back exercises illustrated in a pamphlet. All such patient–physician transactions that are impacted by cost considerations, where the patient wants the more costly strategy, are key areas for negotiation.

The Clash between the Values of the Individual and Society

Negotiation as a part of shared decision making also comes up each time there is a clash between what patients want and what society in one or another of its aspects rejects. Terri Jackson, arguing from an Australian perspective, illustrates the clash in terms of the struggles of U.S. citizens with managed care.[15] Jackson refers to the work of political philosopher Michael Walzer in attempting to define the boundaries between *market-based* and *politics-based* spheres of interest.[16] Walzer argues that all societies declare some transactions to be *blocked exchanges,* exchanges that are illegitimate to transact in the marketplace. An example of such a *blocked exchange* in Western culture is the prohibition on selling jury votes. Jackson believes, after listening to Americans talk about health care, that health care in the United States is also a blocked exchange that Americans are reluctant to allow to be transacted through a market: "Never mind that the care may not be effective, or may even be harmful—the underlying argument seems to be that patient access to care should not be restricted in order to maximize insurers' or employer's profits."[17] This represents another area for negotiation as a part of shared decision making.

There are, of course, limits on the information used in such systematic approaches. Two such limitations involve information about the clinical competencies of specific physicians and information about alternatives. These limitations will persist in the patient–physician relationship until the patient begins to ask specific questions. In the case of *physician competency,* the institution would like each of its physicians to be seen as the best physician for the patient. In most decision analyses, there is no explicit identification of the specific competencies of the physician performing the procedure. How should national, state, local, hospital, clinic, and physician competencies and experience levels in conducting a particular procedure be communicated to patients for risk assessment? And who would perform this evaluation of competency at each level? In the case of alternatives, physicians will typically advise what they consider to be the best alternative available within their institution. Should all possible alternatives be provided, or just those alternatives present in the patient's local community? The patient's region of the country? The United States? Internationally?

Consideration of the limitations placed on alternatives itself opens the door to potential emotional consequences in a patient. These emotional issues can be disturbing, not only to the patient, but to his or her family and to significant others. The focus on shared decision making in relation to affect and emotions includes the same information as our scientific approach, but takes on the task of asking additional key questions related to the *consequences* of the information. Consequences of information could relate to the patient's state of mind and health after hearing about the real-world constraints and limitations on alternatives available within a particular medical institution, a region, or country. They could also relate to intrafamilial conflict regarding the alternatives.

The consequences of information are not limitations placed on the patient by the severity of the patient's condition or disease process. Rather, they can be viewed as potentials or perils borne by the patient who wants to engage with a provider in shared decision making. These are the facts of shared decision making as viewed and considered from its many component perspectives.[18]

Where there is less economic restriction on alternatives, there is greater opportunity to engage patients in a choice or selection process. However, where patients have access to and knowledge about a wider range of alternatives than is available to them—for example, in other health care systems or other regions of the country—the emotional aspects of decision making will come to the fore.

Katz's Alternative View of Shared Decision Making

Jay Katz, a psychiatrist, introduced the idea of shared decision making between doctors and patients as an alternative to the informed consent doctrine expounded by Judge Robinson in *Canterbury v. Spence*. He believes that physicians need to learn to differentiate between a patient's acquiescence to a physician's decision and the patient's consent.[19] Katz asserts that

> the impetus for change in traditional patterns of communication between doctors and patients came not from medicine but from law. In a 1957 California case [*Salgo v. Stanford Board of Trustees*], and a 1960 Kansas case [*Natanson v. Kline*], judges were astounded and troubled by these undisputed facts: That without any disclosure of risks, new technologies had been employed which promised great benefits but exposed patients to formidable and uncontrolled harm. In the California case, a patient suffered a permanent paralysis of his lower extremities subsequent to the injection of a dye, sodium urokan, to locate a block in the abdominal aorta. In the Kansas case, a patient suffered severe injuries from cobalt radiation administered, instead of conventional x-ray treatment, subsequent to a mastectomy for breast cancer.[20]

For Katz, the introduction of the idea of informed consent into American law in the late 1950s and early 1960s was a shock to physicians: "The emerging legal idea that physicians were from now on obligated to share decision making authority with their patients shocked the medical community, for it constituted a radical break with the silence that had been the hallmark of physician–patient interactions throughout the ages."[21]

However, Katz argues that even after Judge Robinson specified in 1972 that patients had the right to self-decision in medical matters, "judges were still reluctant to face up to implications of their novel doctrine [informed consent], preferring instead to remain quite deferential to the practices of the medical profession." He contends,

> [T]he moral authority of physicians resides in knowing better than others the certainties and the uncertainties that accompany diagnosis, treatment, prognosis, health and disease, as well as the extent and the limits of their *scientific* knowledge and *scientific* ignorance. . . . Physicians must learn to face up to and acknowledge the tragic limitations of their own professional knowledge, their inability to impart all their insights to all patients, and their own personal incapacities—at times more pronounced than others—to devote themselves fully to the needs of their patients. They must learn not to be unduly embarrassed by their personal and professional ignorance and to trust their patients to react appropriately to such acknowledgement.

Katz argues that the view of the patient as a passive person "whose welfare was best protected by their following doctors' orders . . . was dictated by doctors' inability to explain to themselves what was therapeutic and what was not in the practice of medicine," and that "informed consent will remain a fairy tale as long as the idea of joint decision making, based on a commitment to patient autonomy and self-determination, does not become an integral aspect of the ethos of medicine and the law of informed consent."[22]

Katz contends that trust is earned after physicians admit what they know and do not know, first to themselves and then to patients. Sharing uncertainties requires that physicians make changes in their own perspectives.

- Physicians need to be more aware of uncertainties than they commonly are.
- Physicians need to learn how to communicate about uncertainty to their patients.
- Physicians need to shed their embarrassment over acknowledging the true state of uncertainty in their own deliberations and the true state of uncertainties that exist in medicine as an art and as a science.
- Physicians need to acquire a willingness to admit ignorance about benefits and risks.
- Physicians need to acknowledge the existence of alternatives, each with its own known and unknown consequences.
- Physicians need to eschew one single authoritarian recommendation.

- Physicians need to consider carefully how to present uncertainty so that patients will not be overwhelmed by the information they will receive.
- Physicians need to explore the crucial question of how much uncertainty physicians themselves can tolerate without compromising their effectiveness as healers.[23]

Katz describes what he sees as doctors' overemphasis on risk disclosures rather than alternatives. For Katz, the disclosure of alternatives is more crucial to joint decision making. Three points need to be made here.

1. In a profound sense, the disclosure of alternatives and an understanding of alternatives is based on a weighing of risks and benefits by the patient in terms of his or her own value system.
2. At any point in time and relative to the severity of the medical condition, the question of impairments in a patient's cognitive processing of information must be addressed in such a way that it is in fact *the patient* who is making the decision.
3. The patient without any cognitive impairment regarding information and decision making may still want the physician's *medical opinion* in relation to the set of alternatives.

Regarding this third point, there remains the issue of what scientific evidence, clinical experience, or expert consensus the physician is using to construct that medical opinion. The medical opinion should be clear, not just hastily constructed from pure conjecture with no sound basis. The notion of shared decision making that Katz is advocating is based on conversations between physicians and their patients. Katz's early writing emphasizes physician uncertainty regarding the specific aspects of the science and practice of medicine related to the patient's particular condition and disease. But Katz also notes that "ultimately the moral authority of physicians resides in their capacity to sort out *with* patients the choices to be made."[24]

Moving now into research on shared decision making, we begin to see that the attempts to inform patients about uncertainty are still in their developmental stages. Indeed, in the last several years the discussion has shifted from uncertainty to evidence. Today, the notion of *underlying evidence* has become a key issue related to new directions in conversations with patients. Adrian Edwards and Glyn Elwyn raise the question of whether evidence-based patient choice is "inevitable or impossible."[25]

Whereas Katz focuses his interest on conversations between physicians and patients, researchers in evidence-based shared decision making are now investigating how *information providers present constraints on choice*. For example,

Vicki Entwistle and Maire O'Donnell argue that information providers must consider how to present choice based on "policy decisions to restrict the supply of particular interventions [that are] usually based on considerations of research evidence about their relative effectiveness, costs, and available resources."[26] They note that "if options are limited, the question of whether people should be told about all available options or all possible options is a difficult one," and point to the need for "information that explains the *reasons* for constraints on availability." These reasons include "research evidence about lack of effectiveness." The problem becomes one of what is to happen to this information when there is, in fact, active debate regarding what evidence should be used to judge effectiveness.

There is at least a slight shift in focus from Katz's views on uncertainty to Entwistle and O'Donnell's views of evidence-based choice. We can see immediately that there will be issues of how patients are going to react to what are essentially *policy issues* related to constraints based on evidence in the clinical care circumstance. While Katz has focused his attention on the discussions between physicians and their patients in conversations in clinic offices or at the bedside, evidence-based researchers will have to raise and answer the more difficult question of how policy-related constraint questions should be addressed with the patient, especially when there is still heated debate on what should be counted as evidence to judge clinical effectiveness in policy issues. Three questions remain for our evidence-based authors: What type of evidence should be used to ground policy issues? In the absence of such evidence, how will decisions be reached? Finally, how will such information be communicated to a patient facing a constrained set of alternatives?

The issues of evidence and economic constraint are shaped by information. Angela Coulter extends the notion of shared decision making between patients and physicians to that of society and citizenry. In terms of decision making at the societal level, Coulter argues that

> just as decision making at the one-to-one level of the clinical consultation must take account the principles of shared decision making, so must those responsible for allocating health care resources seek legitimacy by involving the public and ensuring that the basis for their decisions is transparent and open to challenge if necessary.[27]

In terms of advertising and marketing, "the potential to increase demand for medical interventions and health care services is seen as a great business opportunity. Whether it will benefit the health of the public depends on the quality of information that will be made available."[28] In fact, this benefit will depend not only on the *quality* of the information, but also on *how* (in what formats) that information is presented to the public. This is the first step to understanding on the part of an individual patient, an individual consumer, or the public at large. *Presentation of information* will be the ground of future battles in medical health care, given the increased access to professional and nonprofessional opinions through the Internet. Information will be crucial to the direction of societal change in terms of medical health care. And it will be more and more important to understand information and to be wary of its motives. Information, whether part of an advertisement or promotion or part of a short message provided by an expert consensus panel, results in similar impacts on the recipient of the message. The issues of information presentation, design, and format can potentially determine the patient or consumer's choice *independent* of content. Patients and consumers need access to all the information they want and to as much help understanding that information as can be provided.

Notes

1. J. Graydon et al., "Information Needs of Women During Early Treatment for Breast Cancer," *Journal of Advanced Nursing* 26, no. 1 (July 1997): 59–64.

2. L. F. Degner et al., "Information Needs and Decisional Preferences in Women with Breast Cancer," *Journal of the American Medical Association* 277, no. 18 (May 14, 1997): 1485–92.

3. S. C. Galloway et al., "Perceived Information Needs and Effect of Symptoms on Activities after Surgery for Lung Cancer," *Canadian Oncology Nursing Journal* 3 (1993): 116–19.

4. See, e.g., I. Mesters et al., "Measuring Information Needs Among Cancer Patients," *Patient Education and Counseling* 43, no. 3 (June 2001): 253–62, p. 257.

5. See E. J. Cassell, A. C. Leon, and S. G. Kaufman, "Preliminary Evidence of Impaired Thinking in Sick Patients," *Annals of Internal Medicine* 134, no. 12 (June 19, 2001): 1120–23.

6. D. K. Ziegler et al., "How Much Information about Adverse Events of Medication Do Patients Want from Physicians?" *Archives of Internal Medicine* 161, no. 5 (March 12, 2001): 706–13.

7. Ziegler et al., "How Much Information," 711.

8. See A. J. Rosoff, *Informed Consent: A Guide for Healthcare Providers* (Rockville, Md.: Aspen Systems, 1981), 163–64.

9. U.S. President's Commission for the Study of Ethical Problems in Medicine and Biomedical and Behavioral Research, "Informed Consent as Active, Shared Decision Making," in *Making Health Care Decisions: A Report on the Ethical and Legal Implications of Informed Consent in the Patient–Practitioner Relationship,* vol. 1 (Washington, D.C.: The Commission, 1982), 15–39.

10. M. R. Tonelli, "The Limits of Evidence-Based Medicine," *Respiratory Care* 46, no. 12 (December 2001): 1435–41.

11. Tonelli, "Limits of Evidence-Based Medicine," 1435.

12. D. Blumenthal et al., "The Duration of Ambulatory Visits to Physicians," *Journal of Family Practice* 48, no. 4 (April 1999): 264–71.

13. For example, see R. M. Veatch, *The Patient–Physician Relationship: The Patient as Partner, Part 2* (Bloomington: Indiana University Press, 1991).

14. D. J. Mazur, *Shared Decision Making in the Patient–Physician Relationship: Challenges for the Patient, Physician, and Medical Institution* (Tampa, Fla.: American College of Physician Executives, 2001), 180–83.

15. T. Jackson, "Rationing versus Rationality: Observations from outside the United States," *Medical Decision Making* 21, no. 4 (July/August 2001): 324–28.

16. M. Walzer, *Sphere of Justice: A Defense of Pluralism and Equality* (Oxford: Basil Blackwell, 1983).

17. Jackson, "Rationing versus Rationality," 326.

18. See E. Glyn and C. Charles, "Shared Decision Making: The Principles and the Competencies," in *Evidence-Based Patient Choice: Inevitable or Impossible?* ed. A. Edwards and G. Elwyn (Oxford: Oxford University Press, 2001), 118–43.

19. J. Katz, "Informed Consent: Must It Remain a Fairy Tale?" *Journal of Contemporary Health Law and Policy* 20 (Spring 1994): 69–91.

20. J. Katz, "Informed Consent," 77; see also Katz, *The Silent World of Doctor and Patient* (New York: Free Press, 1984).

21. Katz, "Informed Consent," 72.

22. Katz, "Informed Consent," 72, 90, 91.

23. Paraphrased from Katz, "Informed Consent," 81–82.

24. Katz, "Informed Consent," 90; italics in original.

25. A. Edwards and G. Elwyn, eds., *Evidence-Based Patient Choice: Inevitable or Impossible?* (Oxford: Oxford University Press, 2001).

26. V. Entwistle and M. O'Donnell, "Evidence-Based Health Care: What Roles for Patients?" in *Evidence-Based Patient Choice: Inevitable or Impossible?* ed. A. Edwards and G. Elwyn (Oxford: Oxford University Press, 2001), 45–46.

27. A. Coulter, "The Future," in *Evidence-Based Patient Choice: Inevitable or Impossible?* ed. A. Edwards and G. Elwyn (Oxford: Oxford University Press, 2001), 318.

28. Coulter, "Future," 310.

18

Decision Support for Patients: It's Here, but What Is It, Why Is It Here, Whom Is It Supposed to Benefit, and Where Is It Going?

WE HAVE SO FAR DISCUSSED DECISION MAKING between patients and physicians. But there is a new idea being vigorously developed in medicine and health care. This is the notion of *decision support* devices and systems. While decision support is also being developed for physician decisions, our sole interest will be in decision support for patients. The term *support,* as we will use it, refers to helping patients better understand and use techniques to improve their decision making. Decision support should involve at least three types of information:

1. Information about the decision in terms of risks, benefits, alternatives, including
   - information about the patient's *chance of survival* with each treatment alternative;
   - information about *quality of life* with respect to each alternative;
   - information about the option *not to intervene at all* or *to delay decision making.*
2. Information about the strategies that can be used to prioritize the above information, including
   - rational decision-making strategies;
   - strategies to aid intuitive decision making;
   - strategies based on different values or value systems.
3. Information about what an individual, medical institution, or health care system can be provided by the use of a decision support device or system, and why.

An important area in the patient–provider relationship that we will not consider under the rubric of *decision support* is the *emotional support* of a patient and family during a difficult decision. Susan McClennan Reece conducted a study of postmenopausal women who had elected to discontinue hormone replacement therapy (HRT), concluding that these women

> do so for a variety of reasons, many of which are connected to the health care system and its providers. Outcomes also suggest that sharing in decision making along with increased education, support, and individualized care is necessary to better address the preventive health care needs of postmenopausal women.[1]

Although I believe that this type of support for patients through difficult decisions is crucial in health care delivery, I do not believe that such support can be adequately discussed or provided within the earlier framework of devices and systems.

John Valusek emphasizes the *decision culture* of the healthcare field:

> The creation of a decision culture that matches the "decision intensity" of the health care field is not a paradigm shift but rather a paradigm addition that properly addresses all aspects of information, from how it is delivered to how it is managed. These changes will take health care beyond its current emphasis on efficient transaction systems to reach safe and effective clinical decision environments, which cannot be achieved with transaction mentalities and processes.[2]

This is the approach to information within which we are considering the use of *decision support devices and systems* in physicians' offices, clinics, hospitals, medical centers, and local, regional, and national health care systems. It is important to note that decision support is also being developed for physician decisions. However, our sole interest will be in decision support for patients.

Why Decision Support by Devices and Systems?

It is less than clear why decision support is being developed now. This question has several aspects:

- Why is decision support by *devices* and *interactive computerized systems* being pursued as a technique or strategy within health care?
- What are the motives of the developers of decision support devices and systems?
- Do patients actually want to interact with such devices and systems?
- What types of devices and systems does the patient believe are most useful?

Ubel, Jepson, and Baron articulate one view on what they term *decision aids:*

> To help patients sort out their treatment preferences, researchers have begun developing a wide range of decision aids. Decision aids go a step beyond educational information by helping patients understand how various treatment choices influence their probability of experiencing specific health outcomes, by providing information about how those outcomes might affect their quality of life, and by structuring information . . . to empower patients to actively participate in their treatment choices.[3]

While the underlying concept of *decision support* should be to *support decisions*, the precise nature of what should be included in any support system is up for grabs.

When we begin to consider the development of *decision support* for patients and individuals being recruited for participation in medical research (e.g., clinical trials involving new drugs), we need to divide this discussion into two areas:

1. The development of *decision support devices*, for example, probability boards, handheld specialized calculators, audiovisual tapes, and mini–software programs. These can potentially be used to support *certain aspects* of a decision, for example, displaying data about survival vs. quality of life, walking patients through a standard gamble, or combining one or two other tools that are used to assess patient preferences.
2. The development of *decision support systems*. Here, there is development of more elaborate computerized interactive systems that expose patients to many more types of information and decision strategies. In addition, such systems allow many more opportunities for patients to increase their understanding of the use of information and of a broader range of strategies for decision making—if the patient is willing to commit the time and effort involved in learning how to interact with the system.

Yet the same issues that face physicians in direct face-to-face conversations also face the developers of decision support devices and systems. I have developed a framework for approaching the evaluation of decision support within a medical institution.[4]

In a sense, decision support for patients is as technically challenging an area as we have so far encountered, and it raises all of the issues we have so far discussed related to scientific information in the new medical conversation and the shared focus of decision making in the multicomponent framework discussed earlier. In particular, these areas of our interest—scientific information and shared decision making—generate two questions: What information is to be included for presentation, and how is that information to be presented in such a fashion that it helps the unique patient interacting with the system? One of the main challenges facing developers of decision support systems is *how to follow through on the information presented.*

The tricky part of decision support for patients is how to address the multiple questions the first-time user will have in terms of *what the system is about.* We have previously developed the concept of *follow-through* of scientific information presented in the media and other venues. The notion of a decision support system implies an understanding of *what it is to make a decision.* And the concept of a decision support system would, by its very nature, require users to think about how they make decisions related to medical health care.

The problem with many of the current decision support systems is that the *interaction time* between the patient and the system is so short that there may be little chance for a systematic presentation of issues. Questions arise in this time-limited context: How much can a patient learn in terms of one session? And how many additional sessions are provided to recheck the thinking that resulted from the first session? But more generally, is there a solid chance of a patient systematically reviewing the types of information and the types of strategies offered by a decision support system in a time-constrained context?

One possible approach to these problems involving time constraint is to develop *subsets* of decision support systems. For a particular medical health decision topic, there would be an introductory module related to how decisions are made in general. Then, a module would introduce the concepts necessary to understand how medical decisions are typically made. One of the next subsets would need to go into the notion of information, characterizing it as *descriptive* or *numerical.* The next subset would acquaint patients with the problems associated with the use of information in decision support systems. These problems would include the interpretation of descriptions and the interpretation of numbers.

What is needed is the development of programs that break down decision making into pieces that can be approached in a more leisurely and less time-intensive manner. In addition, the pieces of the system could be presented in whatever order the patient sees fit. Interesting challenges face the developers of decision support for patients at each step, but particularly when they do not ask patients how they prefer to receive information.

The majority of research on patient preferences and medical decisions has been conducted using *hypothetical scenarios.* Why? One answer is that decision making is so subject to influences that researchers are afraid they might unfairly influence a patient on a major decision. Research has been conducted on both *simple* and *complex scenarios*. A simple scenario, in my view, is one in which a patient is asked to consider one dimension of a problem. An example would be a choice of management strategies in terms of chance of survival in the short vs. the long term. Which of two treatments would a patient choose: the treatment with a better chance of short-term survival—for example, at thirty days—but a worse chance of survival at five years, or a treatment with better long-term chances but a worse short-term forecast? One could also provide data on a patient's chance of medium-term survival, and see how subjects react to this additional element.

A *complex scenario* occurs when the patient is asked to choose between management strategies based on many dimensions. An example might be a woman's decision to embark on hormone replacement therapy (HRT), which involves at least three dimensions: risk of developing heart disease, risk of hip fracture, and risk of developing breast cancer. As noted earlier, Fortin et al. explore the complexity of this issue, trying to assess patient preferences at various time intervals using bar graphs to depict the risk factors. The authors found that "there was no clear consensus about whether graphs should display one disease outcome over many time horizons (time intervals) or many disease outcomes for one time horizon, even though slightly more respondents favored the latter."[5]

The problem facing the developer of a decision support device or system is to decide whether to present all options or just one. Researchers in cognitive psychology have argued that information should be presented in at least two formats.[6] It has also been argued that this dual-formatting can "unbias" the presentation of evidence for the patient.[7] This approach has been tested,[8] but more confirmatory studies are needed.

Whom Is Decision Support Supposed to Benefit?

While decision support devices and interactive computerized systems are in their infancy, the underlying strategies are being used in the area of blood testing for prostate cancer and elsewhere.[9] Here, patients are asked to consider data and then make a decision: Do you want this screening test if you are asymptomatic? The question I would like to pose is the following: Do we know enough about such presentations and effects of data on patients to go beyond the research phase and into local marketing or mass marketing of such devices and decisions? I contend that within this setting, more perspectives need to be developed and heard and more approaches systematically reviewed, interpreted, analyzed, evaluated by patients, and discussed in the medical and scientific literatures. In the meantime, we must do more research on the topics we have so far discussed.

Ideally, decision support should benefit the patient. However, when there is active debate on the usefulness of a diagnostic or screening technique and where cost is involved, I argue that decision support can serve multiple purposes within a medical institution, within localities, within regions, within health care frameworks, and within the nation at large. Decision support can be used for many purposes beyond supporting a consumer's or a patient's decision and can actually detract from optimal decision making. But a computerized system that the patient interacts with on his or her own is far removed from the face-to-face interaction of the patient–physician relationship. Therefore, the purpose of decision support for patients is much harder to assess.

Decision support can be used for the following purposes, many of which may not benefit the patient. The patient should determine what is or is not beneficial in his or her view:

- advertising by medical institutions and health care systems designed to attract individuals as potential new clients;
- providing various types of information to patients, for example, information about
  - particular diagnostic or screening tests offered by the medical institution or health care system,
  - diagnostic or screening tests that the medical institution or health care system would like the patients to *move toward* selecting, or
  - diagnostic or screening tests that the medical institution or health care system would like the patients to *move away from*;
  - particular medical interventions that the medical institution or health care system would like the patients to move toward or away from selecting;
- influencing how consumers and patients value aspects of medical health care in their own lives and in the lives of their families;
  - influencing what decisions patients make within a medical institution or health care system;
  - potentially helping individuals as consumers and patients to become better acquainted with how medical decisions are made in their care and the care of their significant others.

Thus, we see that there are many potential uses for a decision support device or system that may not be related to the way a patient or a rational decision maker would want to be supported.

Where Is It Going?

The question of where decision support is going is still up in the air. Decision support has been tried in research studies and is only now being used within medical institutions. Its challenges are similar to what we have already encountered with the information message in general. Yet it does have key differences that need further explanation. Many times it will be hard to tell what the reasons were for developing the decision support device or system in a particular way. Patients may be influenced toward or away from particular diagnostic or therapeutic strategies in ways that are not overt, but rather employ subtle nuances of wording and subtle shifts of attention in the audio or visual aspects of the system. And so, new developers of support must bear in mind the question of *how* patient decisions should be supported.

Ubel, Jepson, and Baron note that while decision aids often provide *statistical information* and *patient testimonials* to guide treatment choices, the testimonials may overwhelm the statistical information.[10] A research study asking patients their choice between therapies may provide statistical information, but that does not mean the subject will use that information in his or her decision making. We will now consider two studies, one using statistical information and patient testimonials, the other providing patients statistical information but then asking the patient what source of information the patient actually relied on in decision making.

Ubel, Jepson, and Baron have studied the use of statistical information and patient testimonials in decision making.

METHODS: Prospective jurors in Philadelphia County were presented with hypothetical statistical information about the percentage of angina patients who benefit from angioplasty and bypass surgery (50% and 75%, respectively). They were also given written testimonials from hypothetical patients who had benefited or not benefited from each of the two treatments. The numbers of patients benefiting and not benefiting were varied to be either proportionate to the statistical information or disproportionate.

In study 1, all participants received 1 testimonial from a patient who had benefited from angioplasty and 1 from a patient who had not. Participants receiving the proportionate questionnaire version were also given 3 testimonials from patients who benefited from bypass surgery and 1 from a patient who did not, coinciding with the hypothetical statistical information. In contrast, participants receiving the disproportionate questionnaire version received only 1 testimonial from a patient who benefited from surgery and 1 from a patient who did not.

In study 2, all participants received 2 examples of patients who benefited from angioplasty and 2 who did not. Participants with the proportionate questionnaire version received the same testimonials regarding surgery as in study 1.

Those receiving the disproportionate questionnaire version received 2 testimonials from patients who benefited from bypass and 2 from patients who did not.

Finally, a separate set of participants in study 2 received a questionnaire with no testimonials.

RESULTS: In study 1, 30% of participants receiving the disproportionate questionnaire version chose bypass surgery versus 44% of those receiving the proportionate questionnaire (P = 0.002 by χ^2). In study 2, 34% of participants receiving the disproportionate questionnaire version chose bypass surgery versus 37% of those receiving the proportionate questionnaire (P = 0.59 by χ^2). Of those receiving no patient testimonials, 58% chose bypass surgery.

The authors conclude that "the inclusion of written patient testimonials significantly influenced hypothetical treatment choices. Efforts to make the mix of positive versus negative testimonials proportionate to statistical information may, under some circumstances, affect choices in ways that cannot automatically be assumed to be optimal."[11]

Regarding the study sample, Armstrong et al. note that "although the juror pool is highly representative of the population of the city of Philadelphia, it is less representative of other segments of the U.S. population."[12] Nonetheless, it appears that subtleties of phrasing do not stop with considerations of verbal text displayed on computer screens. What if patients request testimonials from other patients and use this information sometimes in place of statistical information? This can lead to the systematic manipulation of information presented to patients as well as manipulation of patient choices. Currently, these influences on patient choice are more demonstrable in the research arena, but as decision aids come into use by physicians in practice, in clinics, in medical institutions, and in health care systems, their influence in terms of shifting patient preferences will need systematic attention and control.

In another study, my colleague and I investigated 140 male patients' preferences for management of localized prostate cancer.[13] While we did not use testimonials, we provided statistical information on two management options, surgery and watchful waiting (expectant management or a wait-and-see approach), and then asked patients whether they based their responses on the data we provided or on some other information source.

Of the 74 patients who preferred surgery, 92 percent reported that their choice "was most influenced by the statement in the surgical explanation that 'It is possible that all the cancer will be removed.'" Of the 59 patients who preferred watchful waiting, 80 percent reported that "the described surgical complications most influenced their decision in favor of expectant management." Interestingly, of the patients preferring watchful waiting, 16 percent reported that they based their choice not on any of the statistical information presented but rather on their own personal experiences (8 percent) or on the experience of family, relatives, or friends (8 percent). Of the patients preferring surgical excision, 6 percent reported basing their choice on the experience of family, relatives, or friends (5 percent) or on their own personal experience (1 percent). Thus, even though the physician or researcher may want patients to base their choices as to medical diagnosis or management on statistical information, patients may still base their choices on individual cases.

If testimonials are used, decision support developers must ask themselves where they will get the testimonials from, and whom they will get as actors. If the budget of the decision support project is low, how will that affect the quality of the testimonials displayed to patients? Will the actors be able to portray the testimonials with equal skill, or will one appear more pleasant or sincere, such that the patient may be more drawn to that individual's diagnostic or therapeutic intervention? Furthermore, how will the patient react to the fact that actors will be portraying patients in the testimonials? Will the acting out of a patient case be considered unethical by some patients, but acceptable by others? These questions will need to be addressed by the entire decision support project.

How Should Patient Decisions Be Supported?

The question of how decisions should be supported can be answered in a number of ways, two of which are by asking the patient and by asking the medical institution. Patients might respond with a number of different answers. First, they might want to be supported based on their current beliefs and understanding. Second, they might like to see the full range of how other patients make decisions. Third, they might like to see how rational decision makers make decisions. Fourth, they might like to see how their physician would make the same decision in their situation. Finally, patients could relinquish all decision making and simply track the physician or significant others they assign to make decisions on their behalf. Developers of decision support, however, must educate patients about how they can be influenced and their decision making manipulated by the way information is presented. Simply asking patients what they want can be potentially problematic if the patients do not recognize what constitutes *unfair influence* in how data are presented.

Alternately, one could ask the directors or chief executive officers of the medical institution and others who have developed the decision support system to explore the reasons that underlie their decision to put money into the development of the device.

In the earlier example, I discussed information conveyed to patients in terms of *survival* and *mortality* at the same time. Such dual presentation is not typical. For years, patients usually received *survival data* from their physicians, under the assumption that a chance for increased survival over time was what patients wanted. McNeil, Weichselbaum, and Pauker demonstrated that certain patients were more interested in the quality of their lives than in their survival.[14] Their original work focused on *top-down* decision making, with *numbers* displayed in isolation. Such research makes assumptions and uses a specific data format, then describes the choices patients make within that format. My colleagues and I have focused on *graphic displays of data*, and on how patients understand and use information from the *bottom up*. That is, we focus on understanding how the presentation of data influences decision making. One needs both approaches to improve decision support systems.

The problem with the use of decision support devices is that they may make assumptions and display segments of information in one way, not in many ways. Thus, patients are not being exposed to the full range of potential decision making. Rather, they are exposed to only segments of information and asked to react. More research is needed in breaking down complex decision making into its components, so that patients can learn at early ages and throughout their medical lives what it is to make a medical decision from an informed position and a fuller perspective. This is the promise of decision support systems. But the task is large.

The focus of decision support on presenting patients outcomes, like survival and quality of life, assumes that patients are interested only in this type of information. But, there are other questions a patient may have:

- What are the *causes* of the adverse outcome that may occur, and is everything being done to minimize the chance of adverse outcomes?
- What is the competency and range of experience of the physicians involved in care?
- How does the provider's institution rank (within the community, locality, region, and nation) in terms of the expertise of its providers and the diagnostic and therapeutic techniques offered?
- How are issues of cost constraining the alternatives?
- Will my medical data be kept safely, securely, and confidentially? What are the limits of the protection provided?

The question for future developers of decision support for patients is, how are these elements to be incorporated into decision support devices and systems?

Additional questions may arise from the standpoint of the human subject of clinical research:

- How confidential is information about me as a human subject being kept? Who has access to it and why is this access allowed?

- How am I being identified in different research studies? Are my data labeled with any unique identifier, such as a name or social security number? If so, why?
- If I donate my tissues and organs for the purposes of clinical research related to genetics, and these tissues and organs are uniquely identified as mine, can this donation hurt me or my family and its future generations at a later time—for example, by making employment or insurance harder to obtain?

Elwyn et al. have argued that the promise of systematic approaches to decision making will not be met because of time and economic constraints in the clinic visit.[15] Steven Woolf points to the nature of the logical arguments made in this setting when compelling data are absent: "Nuances in shared decision making materials (scripted presentations, charts, pamphlets, videotapes, and software) can subtly alter the appearance of the tradeoffs, but a reference standard for the 'truth' is often lacking."[16] Woolf illustrates the point with reference to Volk and Spann's finding that men viewing shared decision making materials were 35 percent less likely than others to opt for prostate-specific antigen (PSA) testing: "This finding indicates to proponents of shared decision making an example of patients making more informed choices, whereas to proponents of screening it suggests the possibility of bias in the presentations." [17]

Defining what is clear and compelling information, and how it can be presented for understanding and use in decision making, is the clear challenge of the new millennium. Yet the interface among three areas key to the new medical conversation—decision support, the short information message, and the peer-reviewed medical literature—still needs to be sharpened.

Decision Support and the Impacts of the Short Information Message and Messy Scientific Literature

Decision support cannot be considered in isolation. Particularly when it is focusing on statistical information, it needs to be viewed in light of other sources of information to which patients have access. Notable among these are short information messages involving clinical alerts. Clinical alerts appear in the media prior to the publication of a scientific article, and well before there is a full discussion of its scientific and statistical points. These clinical alerts take on specific significance when they are directed not only to patients and the media but also to clinicians.

Gross et al. studied prepublication release of carotid endarterectomy (CEA) trial results—via National Institutes of Health Clinical Alerts—to determine whether these alerts prompt changes in patient care that are consistent with the new medical evidence contained in the alert.[18] They studied longitudinal data about hospital discharges from the Healthcare Cost and Utilization Project for patients who had carotid endarterectomy performed in acute care hospitals in seven states (New York, California, Pennsylvania, Florida, Colorado, Illinois, and Wisconsin). Gross et al. found that "prepublication dissemination of CEA trial results with clinical alerts was associated with prompt and substantial changes in medical practice," but that "the observed changes suggest that the results were extrapolated to patients and settings not directly supported by the trials." Thus, even practicing physicians' decision making and behaviors may be significantly and perhaps inappropriately influenced by the short information messages contained in clinical alerts. Yet while an individual practitioner may have to make decisions when there are no solid data, the question for developers of decision support is how, if at all, to integrate such clinical alerts with the decision support device or system being developed?

Our previous discussions of the decisions men face when considering whether to agree to a prostate-specific antigen (PSA) blood test based on an incomplete scientific understanding of the management of prostate cancer mirrors the question a woman faces when considering hormone replacement therapy (HRT). To date, investigators have often been unable to compare the results of one HRT study with another. For example, some studies may ask women whether they are current users of HRT, without defining what is meant by *current*. Is a current user of HRT someone who is on the medication for one day, thirty days, a year, or longer? Without such definition, how can an investigator compare data from a study using the phrases "current user," "past user," and "never used" with data from a study asking patients, " have you ever used HRT?"

Furthermore, how are the developers of decision support devices and systems going to get around the difficulty of comparing and contrasting the short information messages found in clinical alerts and scientific literature, when these lack a consistent and coherent definition of terms? The questions regarding clinical alerts and what is to be done with a messy scientific literature need to be faced on a day-to-day basis as part of the ongoing conversation of patients and providers in decision making with patients. In addition, they need to be addressed in the dialogue that should be taking place among patients, providers, and investigators with the developers of decision support devices and systems. These questions are not going to go away. They must be faced head on.

A key challenge for developers of decision support systems is to figure out what needs to be done to realize the promise of a system that provides patients a full and complete discussion of the range of decisions they face and that provides patients with the scientific data they need to know and reflect on in making those decisions. This is the challenge for future generations of students in the fields of science, computer science, decision science, cognitive science, philosophy, sociology, and other disciplines, as they set about learning for their own life's work. I hope this book has allowed you to consider this challenge of decision support systems as one you would like to take up in your own careers in the media, the sciences, clinical medicine, and clinical research.

Notes

1. S. M. Reece, "Weighing the Cons and Pros: Women's Reasons for Discontinuing Hormone Replacement Therapy," *Health Care For Women International* 23, no. 1 (January 2002): 19–32, quoted from p. 19.

2. J. R. Valusek, "Decision Support: A Paradigm Addition for Patient Safety," *Journal of Healthcare Information Management* 16, no. 1 (Winter 2002): 34–39, quoted from p. 34.

3. P. A. Ubel, C. Jepson, and J. Baron, "The Inclusion of Patient Testimonials in Decision Aids: Effects on Treatment Choices," *Medical Decision Making* 21, no. 1 (January/February 2001): 60–68, quoted from p. 60.

4. D. J. Mazur, *Shared Decision Making in the Patient–Physician Relationship: Challenges Facing Patients, Physicians, and Medical Institutions* (Tampa, Fla.: American College of Physician Executives, 2001).

5. J. M. Fortin et al., "Identifying Patient Preferences for Communicating Risk Estimates: A Descriptive Pilot Study," *BMC Medical Informatics and Decision Making* 1, no. 1, at www.biomedcentral.com (accessed July 3, 2002).

6. See D. A. Redelmeir, P. Rozin, and D. Kahneman, "Understanding Patients' Decisions: Cognitive and Emotional Perspectives," *Journal of the American Medical Association* 270, no. 1 (July 7, 1993): 72–76; and D. A. Redelmeier and A. Tversky, "On the Framing of Multiple Prospects," *Psychological Science* 3 (1992): 191–93.

7. K. Armstrong et al., "Effect of Framing as Gain Versus Loss on Understanding and Hypothetical Treatment Choices: Survival and Mortality Curves," *Medical Decision Making* 22, no. 1 (January/February 2002): 76–83.

8. D. J. Mazur and D. H. Hickam, "Treatment Preferences of Patients and Physicians: Influences of Summary Data When Framing Effects Are Controlled," *Medical Decision Making* 10, no. 1 (January/March 1990): 2–5.

9. S. H. Woolf and S. F. Rothemich, "Screening for Prostate Cancer: The Roles of Science, Policy, and Opinion in Determining What Is Best for Patients," *Annual Review of Medicine* 50 (1999): 207–21.

10. Ubel, Jepson, and Baron, "Inclusion of Patient Testimonials."

11. Ubel, Jepson, and Baron, "Inclusion of Patient Testimonials," 60.

12. Armstrong et al., "Effect of Framing as Gain Versus Loss," 76–83, quoted from p. 83.

13. D. J. Mazur and D. H. Hickam, "Patient Preferences for Management of Localized Prostate Cancer," *Western Journal of Medicine* 165, nos. 1–2 (July/August 1996): 26–30.

14. B. J. McNeil, R. Weichselbaum, and S. G. Pauker, "Speech and Survival: Tradeoffs Between Quality and Quantity of Life in Laryngeal Cancer," *New England Journal of Medicine* 305, no. 17 (October 22, 1981): 982–87.

15. G. Elwyn et al., "Decision Analysis in Patient Care," *Lancet* 358 (August 18, 2001): 571–74.

16. S. H. Woolf, "The Logic and Limits of Shared Decision Making," *Journal of Urology* 166, no. 1 (July 2001): 244–45, quoted from p. 224.

17. Woolf, "Logic and Limits," 244; see R. J. Volk and S. J. Spann, "Decision-Aids for Prostate Cancer Screening," *Journal of Family Practice* 49, no. 5 (May 2000): 425–27.

18. C. P. Gross et al., "Relation between Prepublication Release of Clinical Trial Results and the Practice of Carotid Endarterectomy," *Journal of the American Medical Association* 284, no. 22 (December 13, 2000): 2886–93.

19

Summary and Conclusions

WE CAN NOW SEE THAT OUR JOURNEY from advertisement to shared decision making and decision support has not been a solid or a steady evolution, as a reader first approaching this topic may have assumed. At one extreme, short informational messages continue to surround us on a daily basis, and they span a range from advertisement to education. These messages introduce patients to glimpses of ideas about what they should be considering as medical health alternatives in their own lives. Some of these short information messages may be purely promotional, without any attempt at education. Others may give a consumer or patient only a glimpse of what is going on. At the other extreme, I am concerned that the current emphasis on decision support is as problematic as the short information message.

Although we may talk about shared decision making in medical care and the provider–patient relationship, the original goal of the decision scientist, to provide a *full discussion* of risks and benefits among a *full set* of alternatives, may or may not be attainable. And it is far from clear whether patients actually want to participate in such a fully shared decision-making environment. Decision support systems may be very useful in providing those patients who want to understand how better decisions can be made with the tools used in the more formal processes of the decision scientist. The task at hand is introducing decision support to patients in the best possible ways, to see if patients want to use it.

The comparative strategies of the decision scientist may be of interest to some patients, but will the majority want to pay the price in time and energy? While a decision support device may be easier to use, much in terms of weighing and balancing a patient's preferences (risk vs. benefit, survival vs. quality of life) will be abbreviated and potentially hidden.

Here, much more work needs to be done to find out what kind of decision making patients want in their lives. Which patients are more interested in long-term survival, and how can we best support those aims? Which patients are more interested in the quality of their lives in the short and medium terms, and how can we best guide them in their decisions? Which patients do not know what their interests are, and how can we approach this situation? Which patients now want to shift from survival considerations to quality-of-life considerations, or vice versa, and how can we accommodate this shift?

The exploration of ideas we have undertaken extends from a lower level of advertisement or promotion of products to a much fuller exposition of information in shared decision making and decision support. However, it must be recognized that the systematic discussion of information we have attempted continues on a daily basis in many environments—clinic, medical center, computer science laboratories, cognitive psychology laboratories, and so forth—and we have a long way to go before we see clear answers that can help patients with their actual decision making in clinical medicine.

The notion of what constitutes shared decision making is still developing. Whether shared decision making should focus solely on provision of information to patients is still in question, as is the very definition of what is a *better* decision from the patient's perspective—as a human being, as a consumer of medical and health care, and as a citizen in society. We have viewed the evolution of various concepts in shared decision making as they have so far emerged in the published medical literature. We can readily see that there is still a long way to go in their development and implementation, given the increasing constraints on the relationship between patients and their providers. I hope this book may serve in some way to invite the patient into this debate, which relates directly to the care we, our loved ones, and all of us receive in the course of our lives. But most of all, I hope this debate has in some way enticed the undergraduate and graduate reader to become interested in these issues as part of their lives' work in the sciences, the social sciences, philosophy, communications, journalism, or other studies. For unless we have the continued interest of the young in the generation and development of questions related to *how to best communicate information*, the consumer and patient will always be at a loss in an imperfect, partial system.

Index

advertising: advertising message vs. information message, xiii; consumer advertising, broadcast, 44; vs. educating, different aims, 5–6; "hook" as inappropriate, unfair, deceptive, 6–14; increasing demand for medical services, 160; using "science" to set the "hook," 5–6. *See also* direct-to-consumer advertising; ethics; new medical information

autonomy: and benevolence, 37; self-decision, ix; self-determination, 32, 37

Belmont Report, 39, 71–72. *See also* informed consent (clinical research)

bias (cognitive bias), 115; availability bias, 38; cognitive processing, 118–119; hindsight bias, 69

breast cancer, 9–11

CAM (complementary alternative medicine), 43, 49

choice. *See* rationality

cognitive bias. *See* bias

communication: clear communication, xi–xii; communication research, 151–51; flow of information, 25–26; free market communication, 40; non-scientific language, xi, 18, 22–26, 152; pieces of science subsumed in advertising, 40. *See also* information

communications engineering: approaches to information, 63–64; pragmatic needs of government and engineering, 62. *See also* risk communication

competence/competency: of adult patient, 69–70; capacity vs. competency, 32–37; decisionmaking capacity (decisional capacity), 19–20; 36; global understanding, 152; human memory and recall of information, 38–39; impact of mood (depression) on, 35; rational choice, 34; rational decisional capacity, 34; standards for competency, 70; and types of information, 32–34; verbatim memories vs. gist memories, 39. *See also* bias; rationality

comprehension. *See* understanding

contracts (legal contracts): vs. fiduciary duties, 32, 74. *See also* informed

About the Author

Dennis Mazur is professor of medicine at the Oregon Health and Science University, and chairman of the Institutional Review Board of the Department of Veterans Affairs Medical Center in Portland, Oregon. He is feature editor of the journal *Medical Decision Making*, where his column, "Recent Developments in Law and Ethics," is a regular feature. Dr. Mazur serves on the Multifactorial Assessment Plan (MAP) focus group of the Office of Research Compliance and Assurance (ORCA) at the Department of Veterans Affairs Medical Center in Washington, D.C.; he is a consultant to the Veterans Integrated Service Network (VISN 20) of the Department of Veterans Affairs on issues related to medical scientific research and the protection of human subjects; and he is course director for Legal, Regulatory, and Ethical Aspects of Clinical Trials with the Human Investigations Program (HIP) of the Oregon Health and Science University.

Dr. Mazur serves as a reviewer of manuscripts for major peer-reviewed medical journals in the United States, Canada, and Europe, including the *Journal of the American Medical Association*, the *New England Journal of Medicine*, the *Journal of the National Cancer Institute*, the *Canadian Medical Association Journal*, the *Annals of Internal Medicine*, and *The Lancet*.